A Mathematical Structure for Emergent Computation

Nonconvex Optimization and Its Applications

Volume 36

The titles published in this series are listed at the end of this volume.

A Mathematical Structure for Emergent Computation

by

Victor Korotkich
Faculty of Informatics and Communication,
Central Queensland University,
Mackay, Queensland, Australia

SPRINGER-SCIENCE+BUSINESS MEDIA, B.V.

A C.I.P. Catalogue record for this book is available from the Library of Congress.

ISBN 978-1-4613-7424-4 ISBN 978-1-4615-5313-7 (eBook)
DOI 10.1007/978-1-4615-5313-7

Printed on acid-free paper

Contents

Preface

The world is abundant with emergent computation. It can often be viewed when seemingly separate components combine and form something new, in which they act harmoniously together. Depending on the context, the components may represent cells in the human immune system, termites in a colony nest, individual atoms in a molecule, or any of a wide variety of other examples (for example, [1]).

Research in emergent computation is huge and interdisciplinary. It includes an ever growing number of disciplines such as computer science, mathematics, physics, cellular and molecular biology, social and economical sciences with each of them making a valuable concrete contribution. Increasingly, separate disciplines are indicating that at some deep level there are certain universal principles of emergent computation. In particular, the possible existence of these principles is the main motive underlying systems, complexity and chaos theories (for example, [2], [3], [4] and references therein) and their exploration constitutes research directions in many organizations such as the Santa Fe Institute [5].

It is becoming more and more evident that an understanding of the universal principles requires a profound revision of the most fundamental ideas (for example, [6], [7] and references therein). Probably the most noticeable attempt of such a rethinking is a world view in which natural systems are seen not as separate entities but as integrated parts of a unified whole (for example, [8]). In Cartesian world view it is assumed that natural systems are independently existing parts interacting in spatio-temporal void, and this is believed to explain everything.

The supporting evidence for the new world view comes mainly from quantum physics (for example, [8], [9] and references therein). In particular, the conception of the world as an interconnected web of relations is one of the major themes that recur throughout modern physics. The notion of a basic quantum interconnectedness was recognized by Bohr [10] and Heisenberg [11]. The idea of nonlocality can be inferred from Bell's theorem [12] and Bohm provided insight into the meaning of nonlocality by introducing the implicate order [8].

Comprehending the world as a unified whole crucially requires a mathematical structure that can disclose and specify this world view (for example, [6], [7]).

About ten years ago the author was involved in the development and implementation of global optimization algorithms [13]. He started to study questions concerning universal principles of emergent computation as a reaction to the general realization that the NP-complete problem probably could not be practically solved by using the Turing model of computation.

At first, results of these studies were published as separate fragments without any reference to a mathematical structure, which might underlie the principles of emergent computation [14], [15], [16], [17]. Only during the last five years have the studies been crystallising towards such a mathematical structure as outlined in the series of publications [18], [19], [20] [21], [22], [23]. This book summarises these studies, which fall into three main categories: development of the mathematical structure, probing universal principles of emergent computation and a model of computation relevant to those universal principles.

Development of the mathematical structure. It was gradually realized during the studies that the problem of proposing such a structure was mainly about finding proper irreducible concepts to base the structure on and how to develop it without ever once referring to something that is not irreducible. A mathematical structure that has this property is termed a final theory [24]. In particular, no physical concepts can be used as there is no final theory in physics to date.

The structure presented in the book as a mathematical description of the unified whole is based on integers as the single concept and is a set of integer relations organized in a special way. This structure is called a web of relations. The principal radical power of the web of relations comes from a new meaning for the nature of integer relation.

Integer and integer relation are an integral part of our mental equipment, but the scope of humans to fully appreciate integer relation is quite limited. Integer relation is but a set of abstract symbols subject to operational rules of arithmetic. Usually, integer relations appear as solutions to Diophantine equations and questions about their existence and character are of main interest. Probably the most well-known example of this type of understanding of integer relations is the last Fermat's theorem. However, integer relations do not appear to us as rooted in firm reality and have the power to evoke in our minds images of concrete objects.

In the book the scope of integer relation is extended and a new perception of it as a geometrical object is developed. In particular, it is shown that integer relations have a dual existence: on one hand they are, of course, of arithmetic character; on the other hand they find a complete "incarnation" as two-dimensional geometrical objects that form into each other. This dual character of integer relations gives advantages to the web of relations as it creates a new visible world of integer relations, in which logic and intuition come together.

The web of relations is nothing but an integrated collection of hierarchical formations. This produces more and more complex elements as compositions of elements from the lower level. The nature of the hierarchical formations is very close to what is associated with complexity. In particular, the Oxford Dictionary defines something as "complex" if it is "made of (usually several) closely connected parts" or intuitively it is understood that in order to have a complex two or more components are needed, which are joined in such a way that it is difficult to separate them.

There is a substantial literature of attempts to develop a theory of complexity (for example, [25], [26], [27], [28], [29] and references therein). However, in spite of its important role in science and powerful intuitive meanings, complexity does not have generally accepted definitions in rigorous mathematical terms.

The web of relations is proposed as an universal scale to measure the complexity of things in terms of hierarchical formations of integer relations. These hierarchical formations are used to introduce a concept of complexity, called structural complexity, defined as the maximum level of elements that the hierarchical formation produces. The concept is useful in describing the hierarchical formations and plays a very important role.

The web of relations is a final theory with regard to the existence of the elements and their hierarchical formations. It is completely specified by integers, which are considered as the most fundamental entities irreducible to something more simpler. Namely, surprisingly or not ultimate building blocks of the web of relations are integers. They initiate hierarchical formations, level by level, producing more complex elements of the web of relations, which are nothing but integer relations.

Eventually this attempt to capture the unified whole in mathematical terms results in the discovery of a structure, which may be seen as a Platonic ideal, and in essence suggests the idea that the world is a reflection of this undivided, interconnected structure of integer relations.

With the web of relations at hand, clarification is then given in the book as to how it is used in understanding universal principles of emergent computation.

Probing universal principles of emergent computation. The web of relations as a final theory is used to probe universal principles of emergent computation. In particular, a principle is formulated to discuss a variety of issues connected with the mathematical specification of natural systems in terms of hierarchical formations of the web of relations. The principle cannot be derived from the web of relations only and other means are needed for its justification.

However, it is possible to discern outlines of the principle as the hierarchical formations have the preferred bottom-up direction of growth. This suggests that hierarchical formations achieving the maximum level among possible ones stand out as special. This observation accords well with the recognition that the unfolding of ever-greater complexity may be a fundamental property of nature (for example, [30]). This recognition is realized in defining the above principle in a concrete mathematical form by using the concept of structural complexity.

Principle. *A natural system specified in terms of the web of relations comprises hierarchical formations that have maximum possible structural complexity.*

The principle figuratively says that "nature realizes structures with maximum complexity". Since the days of Darwin, evolution has been associated with an increase in complexity. What constitutes the principal difference is that in our case the definition of complexity is formally rigorous and, crucially, is backed up by a final mathematical structure.

A model of computation relevant to those universal principles. To study the principle a model of computation that captures emergent computation and provides abstractions for the development and implementation of optimization algorithms is proposed. The model is based on the El Farol problem suggested by W.B. Arhur [31]. There are several reasons why the El Farol problem is proper for our purposes.

Firstly, the group of $N \geq 2$ people in the El Farol problem exhibits collective behaviour that emerges from their individual decision making rules. The problem easily fits in the framework of emergent computation. Computing the number of seats in the bar is a useful information processing ability of the group. For example, it is easy to appreciate how lucky the owner of the bar would be to have this group of people as customers. Each time for whatever reasons he or she changes the number of the seats, the bar, without any restructuring cost, has exactly as many people as it can accommodate.

Secondly, the behaviour of the group in the El Farol problem can be completely characterized by a matrix consisting of binary sequences. This gives a clear representation in terms of hierarchical formations of the of the web of relations and links with different areas expected to be helpful in future.

Thirdly, the problem has a social aspect which can be tested by the principle. Hidden behind binary sequences of the problem's description, there is a social process involving people and their interaction. This process may be interpreted by the people in categories that involve more moral values rather than formal ones. Participating in the process people may be concerned about equal opportunities, responsibility and other moral issues. But these categories may be also applied to interpret solutions given by the principle. For example, it would be interesting to know if everyone visits the bar regularly or whether they visit according to a preference.

As the principle is not subject to the constraints of the Turing model of computation, a problem of how it can best be algorithmically approximated is considered. The approximation problem turns out to a very difficult one. It is explored as a special case by reducing to the problem of binary sequence prediction with structural complexity as the criterion. An optimal algorithm to the last problem is found. The algorithm gives an experimental verification of the principle by a connection with period doubling, an observable phenomenon of self-organization in nature.

The chapters of the book specify its contents.

The aim of Chapter 1 is to present an integer code series (ICS), which is our main instrument in the discovery of the mathematical structure. A rigorous derivation of the integer code series is presented in section 1 and its coefficients are given explicitly in section 2 of the chapter.

The integer code series gives an interesting representation for definite integrals of piecewise constant functions by combining geometry and integers. Geometry comes in the combination as piecewise constant functions admit visualization in terms of the graphs. Integers come in the combination as the representation includes them and their powers as ingredients. By this combination ICS makes it possible to use visualization to study properties of integers by simply seeing them in graphs of functions.

In section 3 of Chapter 1 to illustrate this potential, it is shown how ICS can be used to make Pythagorean triples appear in a completely different context.

The notion of sequence naturally arises from Cartesian order and plays a very important role in the study of natural systems. In Chapter 2 the integer code series is used to present a global description of sequences as the next step towards the definition of the structure. Namely, given in section 1 this global description of sequences produces in section 2 of the chapter systems of integer relations. These systems of integer relations become the main ingredient in the definition of the structure. The key idea concerning the systems is that they represent a new type of hierarchical formation of integer relations and two-dimensional geometrical patterns.

In Chapter 2 first steps are made to consider these hierarchical formations. As the nature of the hierarchical formations turns out to be very close to what is associated with complexity, they are proposed to measure the complexity of things. For this purpose a concept of complexity, called structural complexity, is introduced. The concept is presented and illustrated in section 3. It is useful in describing the hierarchical formations and plays a very important role.

In section 4 of Chapter 2 it is shown that the Prouhet-Thue-Morse (PTM) sequence has a remarkable property in terms of structural complexity. By using period doubling the property connects structural complexity with self-organization and supports the concept to be used as a measure of complexity. In section 5 of the chapter two-dimensional geometrical patterns, called integer patterns, unfolding the systems of integer relations into visible images are presented. This prepares us to perceive integer relations as geometrical objects that can form into each other.

In section 6 of Chapter 2 hierarchical formations for a particular system of integer relations are visualized. This gives an interesting picture containing nonlocal order and large symmetry. The picture hopefully gives us insight into the harmony of the structure before its definition.

The example in section 6 helps us to understand that the hierarchical formation constitutes one possible hierarchical formation among many others admitted by a larger structure.

In Chapter 3 this structure is considered and defined. This consideration consists of two parts. Firstly, in section 1 the system of integer relation is associated with a hierarchical formation of integer relations. Secondly, in section 2 an isomorphism between the hierarchical formation of integer relations and a hierarchical formation of integer patterns is shown. This helps us in section 3 of Chapter 3 to present a structure, called a web of relations, that integrates all such hierarchical formations into one whole. In section 4 of the chapter we argue in favour of the web of relations to be considered as a final theory. The web of relations does not rely on information about space-time and is based on integers as the single assumption.

In Chapter 4 the web of relations as a final theory is used to probe universal principles of emergent computation. In section 1 of the chapter a principle is presented to discuss a variety of issues connected with the mathematical specification of natural systems in terms of hierarchical formations of the web of relations. The principle cannot be derived from the web of relations only and other means are needed for its justification. However, it is possible to discern outlines of the principle as the hierarchical formations have the preferred bottom-up direction of growth. This suggests that hierarchical formations achieving the maximum level among possible ones stand out as special. This observation accords well with the recognition that the unfolding of ever-greater complexity may be a fundamental property of nature. This recognition is realized in defining the above principle in a concrete mathematical form by using the concept of structural complexity.

In section 2 of Chapter 4 to study the principle a model that captures emergent computation and provides abstractions for the development of optimization algorithms is considered. In the model the thinking of decision makers is an integral part of the situation in which they participate. This limits the use of the standard approach in which natural systems are viewed as independent and separable. In section 3 of the chapter by using the model the principle is expressed in terms of a notion of coherent solution. The notion says that binary sequences of a coherent solution maximize structural complexity.

In Chapter 5 a first step to find verifications of the principle is made. This step is taken to demonstrate that the principle can be translated into a form that appears in the description of self-organization.

One of the important discoveries apparent from the study of natural systems is that self-organization is closely associated with chaos. Many of the natural systems approach chaos through period doubling. The main result of Chapter 5 is that the principle translated in the conventional context of space-time is related to period doubling, an observable phenomenon of self-ogranization in nature. This result can be seen as an experimental verification of the principle.

In particular, to put the translation of the principle in a proper context some known results concerning the symbolic description of period doubling are recalled in section 1 of Chapter 5. As the principle seems not subject to the constraints of the Turing model of computation, the problem of how coherent solutions can best be algorithmically approximated is presented in section 5.2. The approximation problem turns out to be very difficult. It is considered as a special case by reducing to the problem of binary sequence prediction with structural complexity as the criterion.

An optimal algorithm to the last problem is sought in section 5.3. The idea is to find a description of the optimal algorithm that can be associated with self-organization. As an approximation two descriptions of the optimal algorithm are given. One description associates the optimal algorithm with the symbolic description of period doubling and shows that period doubling is a manifestation of the principle. This result throws light on a full theory connected with the principle as a candidate for universal principles of emergent computation.

Acknowledgments

There are many people to whom I am greatly indebted for their support and attention to this work. I am especially grateful to Jury Panteleev, Igor Erlich, Sergei Tarasov, Michael Potapov, Jury Evtushenko, Vladimir Tsurkov, Graham Wood, Panos Pardalos, Zelda Zabinsky, Robert McDougall, Vladimir Dimitrov, David Ruxton, Denis Cryle, Xing Yu, Russel Stonier and Roderick Jewell. My special thanks to Geoff Danaher for proof-reading the text. Of course, the errors if they remain are my sole responsibility.

I want to thank the granting agencies that supported research leading to the results presented in this book

- the Russian Scientifc Fund through grant N93-012-450,

- the Faculty of Informatics and Communication, Central Queensland University through grant VR VR G 121 A0VK1.

I should express my particular warm thanks to John Martindale, an editor in Kluwer Academic Publishers, for his inspiration, patience, and support. I offer my thanks to his assistant, Sharon Donovan with whom it has been a delight to work. I wish to thank Deborah Doherty, an electronic production editor in Kluwer Academic Publishers, whose support is greatly appreciated in preparing the final camera-ready typescript.

Most of all, my beloved wife Galina, my children Grigori and Maria, are owed very special thanks for their love, patience and understanding, especially when I continually underestimate the time that my writting takes me.

Chapter 1

INTEGER CODE SERIES (ICS)

The aim of this chapter is to present an integer code series (ICS), which is our main instrument in the discovery of the mathematical structure. In particular, ICS is used in the next chapters to develop a new perception of integer relations as geometrical objects and define the structure in a way that is accessible both to logic and intuition. A rigorous derivation of the integer code series is presented in section 1 and its coefficients are given explicitly in section 2.

The integer code series gives an interesting representation for definite integrals of piecewise constant functions by combining geometry and integers. Geometry comes in the combination as piecewise constant functions admit visualization in terms of the graphs. A graph, especially when symmetry is involved, often provides means to see properties of the function, which may be difficult to obtain in the formal way. Integers come in the combination as the representation includes them and their powers as ingredients. By this combination ICS makes it possible to use visualization to study properties of integers by simply seeing them in graphs of functions.

It may be said that the discovery of the structure starts with the observation. When graphs of functions have symmetry, sometimes it is literally possible to see values of definite integrals as they appear to be equal zero. The merit of the integer code series is that it translates this result into integer relations and leads us to conceive them as real geometrical objects that form into each other. In section 3 to illustrate this potential, it is shown how ICS can be used to make Pythagorean triples appear in a completely different context.

—

1. INTEGER CODE SERIES

In this section we present a rigorous derivation of the *integer code series* (ICS), which is our main instrument in the study. It seems that the Taylor series is a heavy disguise for the ICS. It went unnoticed for a long time even though the Taylor series lay at the core of calculus and at the intersection of many fields.

The integer code series was found accidently when observing that zero definite integrals of functions with symmetry led to remarkable relations between integers and their powers [14]. A natural attempt to explore and understand this phenomenon resulted in the discovery of the integer code series [15].

However, it takes a lot of manipulations of the Taylor series, as given in this section, before ICS emerges. That is why the proof of Theorem 1.1 is rather long and very technical. Presently we do not know how to make a short cut to it. The reader, if not interested in technical details, may simply skim section 1 and section 2 and refer directly to section 3 of the chapter. This section illustrates that ICS has a specific potential to study integer relations.

Piecewise constant functions provide a natural representation for *sequences*. The use of this representation and the integer code series leads us in a study of sequences to the structure.

Let $n \geq 2, m$ be integers and $\varepsilon > 0, \delta > 0$ be real numbers.

Definition 1.1 (Piecewise Constant Functions). *Let*

$$W_{\varepsilon\delta}([t_m, t_{m+n}])$$

be a set of piecewise constant functions

$$f : [t_m, t_{m+n}] \to \Re^1$$

such that each function f belonging to the set is constant on

$$(t_{i-1}, t_i], \quad i = m+1, ..., m+n$$

and equals

$$f(t_m) = s_1\delta, \quad f(t) = s_i\delta,$$
$$t \in (t_{i-1}, t_i], \quad i = m+1, ..., m+n,$$

where $t_i = i\varepsilon$, $i = m, ..., m+n$ and s_i, $i = 1, ..., n$ are real numbers.

Since a piecewise constant function of the set $W_{\varepsilon\delta}([t_m, t_{m+n}])$ is completely characterised by a sequence of real numbers, it is useful to make the following

Definition 1.2 (Code of a Function). *Let a function*

$$f \in W_{\varepsilon\delta}([t_m, t_{m+n}])$$

and $s_i\delta$ be its value on the interval

$$(t_{i-1}, t_i], \quad i = m+1, ..., m+n$$

then a sequence $s = s_1...s_n$ is called the code of the function and denoted $c(f)$.

By $f^{[r]}$ denote the rth $r = 1, 2, ...$ integral of a function

$$f \in W_{\varepsilon\delta}([t_m, t_{m+n}]),$$

which is a family of functions whose rth derivative equals f. Also let $f^{[0]} = f$ for later convenience. The notation $f^{[r]}$ in formulas stands for a function of the family implying that values $f^{[i]}(t_m)$, $i = 1, ..., r$ specify the function.

The following theorem gives a formula that completely specifies the rth integral of a function $f \in W_{\varepsilon\delta}([t_m, t_{m+n}])$ in terms of its *code* $c(f)$. In addition, the formula contains *integers* and their *powers* as constituents. These properties of the formula make the integer code series our main instrument in the study. In particular, in Chapter 2, the formula is used to give rise to integer relations, which constitute elements of the structure. For convenience we start the consideration with the case when $m = 0$.

Theorem 1.1 (Integer Code Series). *Let a function*

$$f \in W_{\varepsilon\delta}([t_0, t_n]), \quad c(f) = s_1...s_n$$

and values $f^{[i]}(t_0), i = 1, ..., r$ be known, where $r = 1, 2, ...$. Then the rth integral $f^{[r]}$ of the function f at a point

$$t_{k+1}, \quad k = 0, ..., n-1$$

is completely specified by the code $c(f) = s_1...s_n$ of the function and is given by the formula

$$f^{[r]}(t_{k+1}) = \sum_{i=0}^{r-1} \alpha_{ri}((k+1)^i s_1 + ... + 1^i s_{k+1})\delta\varepsilon^r$$

$$+ \sum_{i=1}^{r} \beta_{r,k+1,i} f^{[i]}(t_0)\varepsilon^{r-i}, \qquad (1.1)$$

where coefficient $\alpha_{ri}, i = 0, ..., r-1$ are independent of the code $c(f)$ and the point t_{k+1}, and coefficients $\beta_{r,k+1,i}, i = 1, ..., r$ are independent of the code $c(f)$.

PROOF. Considering that the first $(r-1)$ derivatives of the rth integral are continuous, we have the following Taylor expansion

$$f^{[r]}(t) = \sum_{i=1}^{r} f^{[i]}(t_k)(t-t_k)^{r-i}/(r-i)!$$

$$+ \int_{t_k}^{t} (t-\eta)^{r-1} f(\eta) d\eta/(r-1)!, \tag{1.2}$$

where $t \in (t_k, t_{k+1}]$. Since $f(\eta) = s_{k+1}$ for all $\eta \in (t_k, t_{k+1}]$ and

$$f(t_k + 0) = s_{k+1}$$

then

$$\int_{t_k}^{t} (t-\eta)^{r-1} f(\eta) d\eta/(r-1)!$$

$$= s_{k+1} \int_{t_k}^{t} (t-\eta)^{r-1} d\eta/(r-1)! = s_{k+1}(t-t_k)^r/r!.$$

Substituting this expression into (1.2) and setting $t = t_{k+1}$ gives

$$f^{[r]}(t_{k+1}) = \sum_{i=1}^{r} f^{[i]}(t_k)\varepsilon^{r-i}/(r-i)! + s_{k+1}\varepsilon^r/r!. \tag{1.3}$$

We exclude $f^{[r]}$ values at all points $t_j, j = 1, ..., k$. Namely, starting from the point $t_k, k = 1, ..., n-1$, we write (1.3) as

$$f^{[r]}(t_{k+1}) = f^{[r]}(t_k) + \sum_{i=1}^{r-1} f^{[i]}(t_k)\varepsilon^{r-i}/(r-i)! + s_{k+1}\varepsilon^r/r!.$$

In its turn for $f^{[r]}(t_k)$ we have

$$f^{[r]}(t_k) = f^{[r]}(t_{k-1}) + \sum_{i=1}^{r-1} f^{[i]}(t_{k-1})\varepsilon^{r-i}/(r-i)! + s_k\varepsilon^r/r!.$$

Similarly, if needed, we continue the excluding procedure by using

$$f^{[r]}(t_j) = f^{[r]}(t_{j-1}) + \sum_{i=1}^{r-1} f^{[i]}(t_{j-1})\varepsilon^{r-i}/(r-i)! + s_j\varepsilon^r/r!$$

as a basic step that connects $f^{[r]}(t_j)$ with $f^{[r]}(t_{j-1}), j = 1, ..., k$. The result of all these steps is

$$f^{[r]}(t_{k+1}) = f^{[r]}(t_0) + \sum_{j=0}^{k}\sum_{i=1}^{r-1} f^{[i]}(t_j)\varepsilon^{r-i}/(r-i)!$$

$$+ \sum_{j=1}^{k+1} s_j \varepsilon^r / r!. \tag{1.4}$$

Next, we consider the proof of the theorem directly and demonstrate it by induction on r.

The base case $r = 1$. Indeed, it is easy to see that

$$f^{[1]}(t_{k+1}) = s_1\delta\varepsilon + ... + s_{k+1}\delta\varepsilon + f^{[1]}(t_0)$$

$$= ((k+1)^0 s_1 + ... + 1^0 s_{k+1})\delta\varepsilon + f^{[1]}(t_0),$$

which proves the claim for $r = 1$.

Induction hypothesis. Let us assume the truth of the theorem for $r-1$ when $r \geq 2$. This means that (1.1) is valid for all $i = 1, ..., r-1$

$$f^{[i]}(t_{k+1}) = \sum_{j=0}^{i-1} \alpha_{ij}((k+1)^j s_1 + ... + 1^j s_{k+1})\delta\varepsilon^i$$

$$+ \sum_{j=1}^{i} \beta_{i,k+1,j} f^{[j]}(t_0)\varepsilon^{i-j},$$

and coefficients $\alpha_{ij}, i = 1, ..., r-1,\ j = 0, ..., i-1$ are independent of the code $c(f)$ and the point t_{k+1}, and coefficients $\beta_{i,k+1,j}, i = 1, ..., r-1,\ j = 1, ..., i$ are independent of the code $c(f)$.

These assumptions given we prove (1.1) for r. As the starting point we take the second term of the right-hand side of (1.4) and consider it as the sum of the contributions

$$\sum_{i=1}^{r-1} f^{[i]}(t_j)\varepsilon^{r-i}/(r-i)! \tag{1.5}$$

from all points $t_j, j = 0, ..., k$.

We focus on (1.5). By the inductive assumption for the contribution of a point $t_j, j = 1, ..., k$ we have

$$f^{[i]}(t_j) = \sum_{l=0}^{i-1} \alpha_{il}(j^l s_1 + ... + 1^l s_j)\delta\varepsilon^i + \sum_{l=1}^{i} \beta_{ijl} f^{[l]}(t_0)\varepsilon^{i-l}. \tag{1.6}$$

Inserting (1.6) into (1.5) gives

$$\sum_{i=1}^{r-1} f^{[i]}(t_j)\varepsilon^{r-i}/(r-i)!$$

$$= \sum_{i=1}^{r-1}\sum_{l=0}^{i-1} \alpha_{il}(j^l s_1 + ... + 1^l s_j)\delta\varepsilon^i/(r-i)!$$

$$+ \sum_{i=1}^{r-1}\sum_{l=1}^{i} \beta_{ijl} f^{[l]}(t_0)\varepsilon^{r-l}/(r-i)!$$

$$= \sum_{i=0}^{r-2} \alpha'_{r-1,i}(j^i s_1 + ... + 1^i s_j)\delta\varepsilon^i + \sum_{l=1}^{r-1} \beta'_{r-1,jl} f^{[l]}(t_0)\varepsilon^{r-l}, \tag{1.7}$$

where

$$\alpha'_{r-1,i} = \sum_{l=1}^{r-1} \alpha_{li}/(r-i-1)!, \quad i = 0, ..., r-1,$$

$$\beta'_{r-1,jl} = \sum_{i=l}^{r-1} \beta_{ijl}/(r-l)!. \tag{1.8}$$

Using (1.7) and summing over all points $t_j, j = 1, ..., k$, we obtain

$$\sum_{j=1}^{k}\sum_{i=1}^{r-1} f^{[i]}(t_j)\varepsilon^{r-i}/(r-i)!$$

$$= \sum_{j=1}^{k}\sum_{i=0}^{r-2} \alpha'_{r-1,i}(j^i s_1 + ... + 1^i s_j)\delta\varepsilon^i$$

$$+ \sum_{j=1}^{k}\sum_{l=1}^{r-1} \beta'_{r-1,j,l} f^{[l]}(t_0)\varepsilon^{r-l}$$

$$= \sum_{i=0}^{r-2} \alpha'_{r-1,i} \sum_{j=1}^{k} s_j \sum_{l=1}^{k+1-j} l^i \delta\varepsilon^r$$

$$\sum_{j=1}^{k}\sum_{l=1}^{r-1} \beta'_{r-1,jl} f^{[l]}(t_0)\varepsilon^{r-l}. \tag{1.9}$$

Note that

$$\sum_{l=1}^{k+1-j} l^i = (\sum_{l=0}^{i} \binom{i+1}{l} (k+1-j+1)^{i+1-l} \mathrm{B}_l)/(i+1),$$

where

$$\binom{i+1}{l} = \frac{(i+1)!}{l!(i+1-l)!}, \quad l = 0, ..., i$$

are the binomial coefficients and $\mathrm{B}_l, l = 0, 1, ...$ are the Bernoulli numbers, we have further for (1.9)

$$\sum_{j=1}^{k} \sum_{i=1}^{r-1} f^{[i]}(t_j) \varepsilon^{r-i}/(r-i)!$$

$$= \sum_{i=0}^{r-2} \alpha'_{r-1,i} \sum_{j=1}^{k} s_j \Big(\sum_{l=0}^{i} \binom{i+1}{l} (k+1-j+1)^{i+1-l} \mathrm{B}_l\Big) \delta\varepsilon^r/(i+1)$$

$$+ \sum_{j=1}^{k} \sum_{l=1}^{r-1} \beta'_{r-1,jl} f^{[l]}(t_0) \varepsilon^{r-l}. \tag{1.10}$$

Define

$$\alpha''_{r-1,i} = \sum_{j=i-1}^{r-1} \alpha'_{r-1,j} \binom{j+1}{j-i+1} \mathrm{B}_{j-i+1}/(j+1), \quad i = 1, ..., r-1,$$

$$\beta''_{r-1,kl} = \sum_{j=1}^{k} \beta'_{r-1,jl}. \tag{1.11}$$

It follows, substituting (1.11) into (1.10), that

$$\sum_{j=1}^{k} \sum_{i=1}^{r-1} f^{[i]}(t_j) \delta\varepsilon^{r-i}/(r-i)!$$

$$= \sum_{i=1}^{r-1} \alpha''_{r-1,i} ((k+1)^i s_1 + ... + 2^i s_k) \delta\varepsilon^r$$

$$+ \sum_{i=1}^{r-1} \beta''_{r-1,k,i} f^{[i]}(t_0) \varepsilon^{r-i}. \tag{1.12}$$

The right-hand side of (1.12) by addition and subtraction of

$$((k+1)s_1 + ... + 2s_k)\varepsilon^r$$

takes the form

$$\sum_{i=1}^{r-1} \alpha''_{r-1,i} ((k+1)^i s_1 + ... + 2^i s_k) \delta\varepsilon^r$$

$$+((k+1)s_1+\dots+2s_k)\delta\varepsilon^r$$

$$-((k+1)s_1+\dots+2s_k)\delta\varepsilon^r+\sum_{i=1}^{r-1}\beta''_{r-1,ki}f^{[i]}(t_0)\varepsilon^{r-i}. \tag{1.13}$$

Define

$$\alpha'''_{r-1,i}=\alpha''_{r-1,i},\quad i=2,\dots,r-1,$$

$$\alpha'''_{r-1,1}=\alpha''_{r-1,1}+1. \tag{1.14}$$

Substituting (1.14) into (1.13), we further modify (1.12)

$$\sum_{j=1}^{k}\sum_{i=1}^{r-1}f^{[i]}(t_j)\delta\varepsilon^{r-i}/(r-i)!$$

$$=\sum_{i=1}^{r-1}\alpha'''_{r-1,i}((k+1)^is_1+\dots+2^is_k)\varepsilon^r$$

$$-((k+1)s_1+\dots+2s_k)\delta\varepsilon^r+\sum_{i=1}^{r-1}\beta''_{r-1,k,i}f^{[i]}(t_0)\varepsilon^{r-i}. \tag{1.15}$$

The following rearrangement of the right-hand side of (1.15) is needed to include s_{k+1} in it. We have

$$\sum_{i=1}^{r-1}\alpha'''_{r-1,i}((k+1)^is_1+((k-1)+1)^is_2+\dots+((2-1)+1)^is_k)\delta\varepsilon^r$$

$$-((k+1)s_1+\dots+2s_k)\delta\varepsilon^r+\sum_{i=1}^{r-1}\beta''_{r-1,k,i}f^{[i]}(t_0)\varepsilon^{r-i}$$

$$=\sum_{i=1}^{r-1}\alpha'''_{r-1,i}\sum_{j=0}^{i}(\binom{i}{j}k^{i-j}s_1+\dots+\binom{i}{j}(2-1)^{i-j}s_k)\varepsilon^r$$

$$-((k+1)s_1+\dots+2s_k)\delta\varepsilon^r+\sum_{i=1}^{r-1}\beta''_{r-1,ki}f^{[i]}(t_0)\varepsilon^{r-i}. \tag{1.16}$$

Let

$$\alpha''''_{r-1,r-i}=\sum_{j=1}^{i}\alpha'''_{r-1,r-j}\binom{r-j}{i-j},\quad i=1,\dots,r. \tag{1.17}$$

Then the right-hand side of (1.16) becomes

$$\sum_{i=0}^{r-1}\alpha''''_{r-1,i}(k^is_1+\dots+1^is_k)\delta\varepsilon^r$$

$$-((k+1)s_1 + ... + 2s_k)\delta\varepsilon^r + \sum_{i=1}^{r-1} \beta''_{r-1,k,i} f^{[i]}(t_0)\varepsilon^{r-i}.$$

Rewriting the last expression in another way and opening the brackets, we have

$$\sum_{i=0}^{r-1} \alpha''''_{r-1,i}((k+1)-1)^i s_1 + ... + (2-1)^i s_k)\delta\varepsilon^r$$

$$-((k+1)s_1 + ... + 2s_k)\delta\varepsilon^r + \sum_{i=1}^{r-1} \beta''_{r-1,k,i} f^{[i]}(t_0)\varepsilon^{r-i}$$

$$= \sum_{i=0}^{r-1} \alpha''''_{r-1,i} \sum_{j=0}^{i}((-1)^j \binom{i}{j}(k+1)^{i-j}s_1 + ... + (-1)^j \binom{i}{j} 2^{i-j} s_k)\delta\varepsilon^r$$

$$-((k+1)s_1 + ... + 2s_k)\varepsilon^r + \sum_{i=1}^{r-1} \beta''_{r-1,k,i} f^{[i]}(t_0)\varepsilon^{r-i}.$$

Using

$$s_{k+1}(1-1)^i = \sum_{j=0}^{i}(-1)^j \binom{i}{j} s_{k+1} = 0, \quad i = 1, ..., r-1,$$

$$s_{k+1} = \sum_{j=0}^{i}(-1)^j \binom{i}{j} s_{k+1}, \quad i = 0,$$

we find

$$\sum_{i=0}^{r-1} \alpha''''_{r-1,i} \sum_{j=0}^{i}((-1)^j \binom{i}{j}(k+1)^{i-j} + ... + (-1)^j \binom{i}{j} 1^{i-j} s_{k+1})\delta\varepsilon^r$$

$$-s_{k+1}\delta\varepsilon^r - ((k+1)s_1 + ... + 2s_k)\delta\varepsilon^r$$

$$+\sum_{i=1}^{r-1} \beta''_{r-1,ki} f^{[i]}(t_0)\delta\varepsilon^{r-i}$$

$$= \sum_{i=0}^{r-1} \alpha''''_{r-1,i} \sum_{j=0}^{i}(-1)^j \binom{i}{j}((k+1)^{i-j} + ... + 2^{i-j}s_k + 1^{i-j}s_{k+1})\delta\varepsilon^r$$

$$-((k+1)s_1 + ... + 2s_k + 1s_{k+1})\delta\varepsilon^r$$

$$+\sum_{i=1}^{r-1} \beta''_{r-1,ki} f^{[i]}(t_0)\delta\varepsilon^{r-i}. \tag{1.18}$$

Let

$$\alpha'''''_{r-1,r-i} = \sum_{j=1}^{i}(-1)^{i-j}\alpha''''_{r-1,r-j}\binom{r-j}{i-j}, \quad i = 1, ..., r. \qquad (1.19)$$

Substitution of this into the right-hand of (1.18) yields

$$\sum_{i=0}^{r-1}\alpha'''''_{r-1,i}((k+1)^i s_1 + ... + 1^i s_{k+1})\delta\varepsilon^r$$

$$-((k+1)s_1 + ... + 1s_{k+1})\delta\varepsilon^r + \sum_{i=1}^{r-1}\beta''_{r-1,k,i}f^{[i]}(t_0)\delta\varepsilon^{r-i}. \qquad (1.20)$$

Since (1.20) is a transformed right-hand side of (1.15), this means

$$\sum_{j=1}^{k}\sum_{i=1}^{r-1} f^{[i]}(t_j)\varepsilon^{r-i}/(r-i)! =$$

$$\sum_{i=0}^{r-1}\alpha'''''_{r-1,i}((k+1)^i s_1 + ... + 1^i s_{k+1})\delta\varepsilon^r$$

$$-((k+1)s_1 + ... + 1s_{k+1})\delta\varepsilon^r + \sum_{i=1}^{r-1}\beta''_{r-1,ki}f^{[i]}(t_0)\varepsilon^{r-i}.$$

In view of the last relation for (1.4) we get

$$f^{[r]}(t_{k+1}) = f^{[r]}(t_0)$$

$$+\sum_{j=0}^{k}\sum_{i=1}^{r-1} f^{[i]}(t_j)\varepsilon^{r-i}/(r-i)! + \sum_{j=1}^{k+1} s_j\delta\varepsilon^r/r!$$

$$= f^{[r]}(t_0) + \sum_{i=1}^{r-1} f^{[i]}(t_0)\varepsilon^{r-i}/(r-i)!$$

$$+\sum_{j=1}^{k}\sum_{i=1}^{r-1} f^{[i]}(t_j)\varepsilon^{r-i}/(r-i)! + \sum_{j=1}^{k+1} s_j\delta\varepsilon^r/r!$$

$$= f^{[r]}(t_0) + \sum_{i=1}^{r-1} f^{[i]}(t_0)\varepsilon^{r-i}/(r-i)!$$

$$+\sum_{i=0}^{r-1}\alpha'''''_{r-1,i}((k+1)^i s_1 + ... + 1^i s_{k+1})\delta\varepsilon^r -$$

$$-((k+1)s_1+...+1s_{k+1})\delta\varepsilon^r$$

$$+\sum_{i=1}^{r-1}\beta''_{r-1,ki}f^{[i]}(t_0)\varepsilon^{r-i}+\sum_{j=1}^{k+1}s_j\delta\varepsilon^r/r!. \tag{1.21}$$

At last we set

$$\alpha_{r0}=\alpha'''''_{r-1,0}+1/r!, \quad \alpha_{r1}=\alpha'''''_{r-1,1}-1,$$

$$\alpha_{ri}=\alpha'''''_{r-1,i}, \quad i=2,...,r-1,$$

$$\beta_{r,k+1,i}=\beta''_{r-1,ki}+1/(r-i)!, \quad i=1,...,r-1,$$

$$\beta_{r,k+1,r}=1. \tag{1.22}$$

Making use of (1.22), we finally come from (1.21) to

$$f^{[r]}(t_{k+1})=\sum_{i=0}^{r-1}\alpha_{ri}((k+1)^i s_1+...+1^i s_{k+1})\delta\varepsilon^r$$

$$+\sum_{i=1}^{r}\beta_{r,k+1,i}f^{[i]}(t_0)\varepsilon^{r-i}.$$

To complete the induction step we prove that coefficients

$$\alpha_{ri}, \quad i=0,...,r-1$$

and

$$\beta_{r,k+1,i}, \quad i=1,...,r$$

satisfy the terms of the theorem.

The proof is an easy consequence of the coefficients' definition. Indeed, let us observe that coefficients $\alpha_{ri}, i=0,...,r-1$ are generated by coefficients $\alpha_{ij}, i=1,...,r-1, \ j=0,...,i-1$ successively in (1.8), (1.11), (1.14), (1.17), (1.19), (1.22). By the inductive assumption

$$\alpha_{ij}, \quad i=1,...,r-1, \quad j=0,...,i-1$$

are independent of the code $c(f)$ and the point t_{k+1}. Also there are no dependences on the code $c(f)$ and the point t_{k+1} in (1.8), (1.11), (1.14), (1.17), (1.19), (1.22). Therefore, coefficients $\alpha_{ri}, i=0,...,r-1$ are independent of the code $c(f)$ and the point t_{k+1}.

Analogously, coefficients $\beta_{r,k+1,i}, i=1,...,r$ are generated by coefficients

$$\beta_{ijl}, \quad i=1,...,r-1, \quad j=1,...,k+1, \quad l=1,...,i$$

successively in (1.8), (1.11), (1.22). Again by the inductive assumption and the character of (1.8), (1.11), (1.22) we have that coefficients $\beta_{r,k+1,i}, i = 1, ..., r$ are independent of the code $c(f)$.

Finally the theorem is proved.

The following result is a generalization of Theorem 1.1.

Theorem 1.2 (Integer Code Series: General Case). *Let a function*

$$f \in W_{\varepsilon\delta}([t_m, t_{m+n}]), \quad c(f) = s_1...s_n$$

and values $f^{[i]}(t_m), i = 1, ..., r$ *be known, where* $r = 1, 2, ...$ *. Then the* r*th integral* $f^{[r]}$ *of the function* f *at a point*

$$t_{m+k+1}, \quad k = 0, ..., n-1$$

is completely specified by the code $c(f) = s_1...s_n$ *of the function and is given by the formula*

$$f^{[r]}(t_{m+k+1}) = \sum_{i=0}^{r-1} \alpha_{rmi}((m+k+1)^i s_1 + ... + (m+1)^i s_{k+1})\varepsilon^r$$

$$+ \sum_{i=1}^{r} \beta_{r,k+1,i} f^{[i]}(t_m)\varepsilon^{r-i}, \qquad (1.23)$$

where coefficients $\alpha_{rmi}, i = 0, ..., r-1$ *are independent of the code* $c(f)$ *and the point* t_{m+k+1}*, and coefficients* $\beta_{r,k+1,i}, i = 1, ..., r$ *are independent of the code* $c(f)$ *and integer* m.

PROOF. Consider a function $g \in W_{\varepsilon\delta}([t_0, t_n])$ such that

$$c(g) = c(f) = s_1...s_n,$$

$$g^{[i]}(t_0) = f^{[i]}(t_m), \quad i = 1, ..., r.$$

By Theorem 1.1 for the function g we have

$$g^{[r]}(t_{k+1}) = \sum_{i=0}^{r-1} \alpha_{ri}((k+1)^i s_1 + ... + 1^i s_{k+1})\delta\varepsilon^r$$

$$+ \sum_{i=1}^{r} \beta_{r,k+1,i} g^{[i]}(t_0)\varepsilon^{r-i}.$$

We rewrite this formula for $g^{[r]}(t_{k+1})$ as

$$g^{[r]}(t_{k+1})$$

$$= \sum_{i=0}^{r-1} \alpha_{ri}(((m+k+1)-m)^i s_1 + \ldots + ((m+1)-m)^i s_{k+1})\varepsilon^r$$

$$+ \sum_{i=1}^{r} \beta_{r,k+1,i} g^{[i]}(t_0)\varepsilon^{r-i}$$

$$= \sum_{i=1}^{r-1} \alpha_{ri} \sum_{j=0}^{i} (-1)^j \binom{i}{j} ((m+k+1)^{i-j} m^j s_1 + \ldots + (m+1)^{i-j} m^j s_{k+1})\delta\varepsilon^r$$

$$-\alpha_{r0}((m+k+1)^0 s_1 + \ldots + (m+1)^0 s_{k+1})\delta\varepsilon^r$$

$$+ \sum_{i=1}^{r} \beta_{r,k+1,i} g^{[i]}(t_0)\varepsilon^{r-i}. \tag{1.24}$$

Let

$$\alpha_{rm,r-i} = \sum_{j=1}^{i} (-1)^j \alpha_{r,r-j} \binom{r-j}{i-j} m^{i-j}, \quad i = 1, \ldots, r, \tag{1.25}$$

where index m is used to identify dependence of coefficients $\alpha_{rmi}, i = 0, \ldots, r-1$ on m. Note that coefficients $\alpha_{rmi}, i = 0, \ldots, r-1$ in (1.25) are independent of the code $c(g)$ and the point t_{k+1}.

Substituting (1.25) into (1.24) yields

$$g^{[r]}(t_{k+1})$$

$$= \sum_{i=0}^{r-1} \alpha_{rmi}((m+k+1)^i s_1 + \ldots + (m+1)^i s_{k+1})\delta\varepsilon^r$$

$$+ \sum_{i=1}^{r} \beta_{r,k+1,i} g^{[i]}(t_0)\varepsilon^{r-i}. \tag{1.26}$$

On the other hand by the construction of the functions, we have

$$f^{[r]}(t') = g^{[r]}(t), \quad t' = t + t_m,$$

where $t \in [t_0, t_n]$, $t' \in [t_m, t_{m+n}]$. This implies that

$$f^{[r]}(t_{m+k+1}) = g^{[r]}(t_{k+1}), \quad k = 0, \ldots, n-1.$$

Consequently, using (1.26) and

$$g^{[i]}(t_0) = f^{[i]}(t_m), \quad i = 1, \ldots, r,$$

we finally obtain (1.23)

$$f^{[r]}(t_{m+k+1})$$

$$= \sum_{i=0}^{r-1} \alpha_{rmi}((m+k+1)^i s_1 + ... + (m+1)^i s_{k+1})\delta\varepsilon^r$$

$$+ \sum_{i=1}^{r} \beta_{r,k+1,i} f^{[i]}(t_m)\varepsilon^{r-i}.$$

This ends the proof of the theorem.

2. COEFFICIENTS OF THE INTEGER CODE SERIES

Theorem 1.2 gives some information about coefficients $\alpha_{rmi}, \beta_{r,k+1,i}$ but it does not specify them explicitly. In this section explicit formulas for the coefficients are derived.

Theorem 1.3 (Coefficients α_{rmi}). *Coefficients α_{rmi} of the integer code series in the formula (1.23) are given by*

$$\alpha_{rmi} = \binom{r}{i}((-1)^{r-i-1}(m+1)^{r-i} + (-1)^{r-i} m^{r-i})/r!, \tag{1.27}$$

where $r = 1, 2, ...,\ i = 0, ..., r-1$ and m is an integer.

PROOF. Consider a function $g \in W_{\varepsilon\delta}([t_m, t_{m+n}])$ such that

$$g(t) = \begin{cases} \delta, & t \in [t_m, t_{m+1}] \\ 0, & t \in [t_{m+1}, t_{m+n}] \end{cases}$$

and $g^{[i]}(t_m) = 0, i = 1, ..., r$.

Observe that the code of the function g is $c(g) = (1, 0, ..., 0)$. Therefore, by Theorem 1.2 for the rth $r = 1, 2, ...$ definite integral of the function at a point $t_{m+k}, k = 1, ..., n$ we have

$$g^{[r]}(t_{m+k}) = \sum_{i=0}^{r-1} \alpha_{rmi}(m+k)^i \delta\varepsilon^r. \tag{1.28}$$

On the other hand, directly r times integrating the function g over the interval $[t_{m+1}, t_{m+n}]$, we get for $t \geq t_{m+1}$

$$g^{[1]}(t) = \delta\varepsilon,$$

$$g^{[2]}(t) = \delta\varepsilon t + a_1,$$

$$\cdot \quad \cdot \quad \cdot \quad \cdot \quad \cdot$$

$$g^{[r]}(t) = \delta\varepsilon t^{r-1}/(r-1)!$$
$$+a_1 t^{r-2}/(r-2)! + ... + a_{r-2}t/1! + a_{r-1}$$
$$= \sum_{i=0}^{r-1} a_{r-i-1} t^i/i!,$$

where $a_0 = \delta\varepsilon$.

Thus, for a point $t_{m+k} = (m+k)\varepsilon,\ k = 1, ..., n$ we obtain

$$g^{[r]}(t_{m+k}) = \sum_{i=0}^{r-1} a_{r-i-1}(m+k)^i \varepsilon^i/i! \tag{1.29}$$

It follows from (1.28) and (1.29) that

$$\sum_{i=0}^{r-1} \alpha_{rmi}(m+k)^i \delta\varepsilon^r = \sum_{i=0}^{r-1} a_{r-i-1}(m+k)^i \varepsilon^i/i!.$$

Consequently,

$$\sum_{i=0}^{r-1} (\alpha_{rmi}\delta\varepsilon^{r-i} - a_{r-i-1}/i!)(m+k)^i \varepsilon^i = 0. \tag{1.30}$$

Note that n can be chosen any positive integer and $k = 1, ..., n$. Therefore, for each $r = 1, 2, ...$ we choose $n = r$ and set (1.30) to be satisfied at r points $t_{m+k}, k = 1, ..., r$. Consequently, the polynomial of degree $r-1$

$$a'_0 t^{r-1} + a'_1 t^{r-2} + ... + a'_{r-1},$$

where

$$a'_i = \alpha_{rmi}\delta\varepsilon^{r-i} - a_{r-i-1}/i!, \quad i = 0, ..., r-1$$

is zero at r points $t_{m+k}, k = 1, ..., r$. This implies that

$$a'_i = 0, \quad i = 0, ..., r-1,$$

since, if a polynomial of degree $r-1$ is zero at r distinct points, then it is identically zero. Hence

$$\alpha_{rmi} = a_{r-i-1}/\varepsilon^{r-i} i!, \quad i = 0, ..., r-1. \tag{1.31}$$

Next, we determine coefficients $a_i, i = 0, ..., r-1$. Denote

$$g(t) = \begin{cases} g_1(t), & t \in [t_m, t_{m+1}] \\ g_2(t), & t \in [t_{m+1}, t_{m+n}] \end{cases}$$

Since all integrals of the function g are continuous, we have for $r = 1, 2, ...$

$$g_1^{[i]}(t_{m+1}) = g_2^{[i]}(t_{m+1}), \quad i = 1, ..., r. \tag{1.32}$$

By using (1.32) we show by induction on $r = 1, 2, ...$ that

$$a_i = \delta\varepsilon^{i+1}((-1)^i(m+1)^{i+1} + (-1)^{1+i}m^{i+1})/(i+1)!, \tag{1.33}$$

where $i = 0, ..., r-1$.

For the base case $r = 1$ we obtain immediately $a_0 = \delta\varepsilon$, which proves the claim for the case.

Assume that the claim (1.33) is valid for r, then we prove it for $r+1$. From (1.29) for $g_2^{[r+1]}$ at the point t_{m+1} we get

$$\begin{aligned} g_2^{[r+1]}(t_{m+1}) &= \sum_{i=0}^{r} a_{r-i}(m+1)^i\varepsilon^i/i! \\ &= \sum_{i=0}^{r} a_i(m+1)^{r-i}\varepsilon^{r-i}/(r-i)! \\ &= a_r + \sum_{i=0}^{r-1} a_i(m+1)^{r-i}\varepsilon^{r-i}/(r-i)!. \end{aligned} \tag{1.34}$$

By using the inductive assumption we insert $a_i, i = 0, ..., r-1$ from (1.33) into (1.34) and obtain

$$\begin{aligned} &\sum_{i=0}^{r-1} \frac{\delta\varepsilon^{r+1}}{(i+1)!(r-i)!}((-1)^i(m+1)^{r+1} + (-1)^{1+i}(m+1)^{r-i}m^{i+1}) \\ &= \delta\varepsilon^{r+1}(m+1)^{r+1}\sum_{i=0}^{r-1}\frac{(-1)^i}{(i+1)!(r-i)!} \\ &+\delta\varepsilon^{r+1}\sum_{i=0}^{r-1}\frac{(-1)^{1+i}(m+1)^{r-i}m^{i+1}}{(i+1)!(r-i)!} \\ &= \delta\varepsilon^{r+1}(m+1)^{r+1}T_1 + \delta\varepsilon^{r+1}T_2. \end{aligned} \tag{1.35}$$

Evaluation of T_1 and T_2 requires more work. Obviously, we have for T_1 when $r' \geq 2$

$$(1-1)^{r'} = \sum_{j=0}^{r'} \frac{(-1)^j r'!}{j!(r'-j)!} = 1 + r'! \sum_{j=1}^{r'} \frac{(-1)^j}{j!(r'-j)!} = 0. \tag{1.36}$$

Letting $j = i+1$ and $r' = r+1$, we further find for (1.36)

$$1 + r'! \sum_{i=0}^{r'-1} \frac{(-1)^{i+1}}{(i+1)!(r'-i-1)!}$$

$$= 1 - (r+1)! \sum_{i=0}^{r} \frac{(-1)^i}{(i+1)!(r-i)!}$$

$$= 1 - (r+1)!\left(\frac{(-1)^r}{(r+1)!} + \sum_{j=0}^{r-1} \frac{(-1)^i}{(i+1)!(r-i)!}\right)$$

$$= 1 + (-1)^{r+1} - (r+1)! T_1 = 0.$$

Hence

$$T_1 = \frac{1 + (-1)^{r+1}}{(r+1)!}. \tag{1.37}$$

It is equally straightforward to get for T_2

$$(m+1-m)^{r'} = \sum_{j=0}^{r'} \frac{(-1)^j r'!(m+1)^{r'-j} m^j}{j!(r'-j)!}$$

$$= (m+1)^{r'} + \sum_{j=1}^{r'} \frac{(-1)^j r'!(m+1)^{r'-j} m^j}{j!(r'-j)!} = 1 \tag{1.38}$$

Similarly, letting $j = i+1$ and $r' = r+1$, we obtain by using (1.38)

$$(m+1)^{r'} + r'! \sum_{i=0}^{r'-1} \frac{(-1)^{i+1}(m+1)^{r'-i-1} m^{i+1}}{(i+1)!(r'-i-1)!}$$

$$= (m+1)^{r+1} + (r+1)! \sum_{i=0}^{r} \frac{(-1)^{i+1}(m+1)^{r-i} m^{i+1}}{(i+1)!(r-i)!}$$

$$= (m+1)^{r+1} + (-1)^{r+1} m^{r+1}$$

$$+ (r+1)! \sum_{i=0}^{r-1} \frac{(-1)^{i+1}(m+1)^{r-i} m^{i+1}}{(i+1)!(r-i)!}$$

$$= (m+1)^{r+1} + (-1)^{r+1}m^{r+1} + (r+1)!T_2 = 1.$$

This gives

$$T_2 = \frac{1-(m+1)^{r+1} - (-1)^{r+1}m^{r+1}}{(r+1)!}. \tag{1.39}$$

Combining (1.37) and (1.39) in (1.35), we further transform (1.34) to get

$$g_2^{[r+1]}(t_{m+1})$$
$$= a_r + \frac{\delta\varepsilon^{r+1}(m+1)^{r+1}}{(r+1)!}(1+(-1)^{r+1})$$
$$+\frac{\delta\varepsilon^{r+1}}{(r+1)!}(1-(m+1)^{r+1} - (-1)^{r+1}m^{r+1}). \tag{1.40}$$

Obviously

$$g_1^{[r+1]}(t_{m+1}) = \frac{\delta\varepsilon^{r+1}}{(r+1)!}. \tag{1.41}$$

From (1.32), by using (1.40) and (1.41), we arrive at

$$a_r = \delta\varepsilon^{r+1}((-1)^r(m+1)^{r+1} + (-1)^{r+1}m^{r+1})/(r+1)!.$$

This completes the induction step.

Lastly, by substituting (1.33) into (1.31), we obtain the desired result

$$\alpha_{rmi} = \binom{r}{i}((-1)^{r-i-1}(m+1)^{r-i} + (-1)^{r-i}m^{r-i})/r!,$$

where $i = 0, ..., r-1$. This proves the theorem.

The next theorem specifies coefficients

$$\beta_{rki}, \quad r = 1, 2, \ldots, \quad k = 1, ..., n, \quad i = 1, ..., r$$

of the integer code series.

Theorem 1.4 (Coefficients β_{rki}). *Coefficients β_{rki} of the integer code series in the formula* (1.23) *are given by*

$$\beta_{rki} = \frac{k^{r-i}}{(r-i)!}, \tag{1.42}$$

where $r = 1, 2, \ldots, k = 1, ..., n, \ i = 1, ..., r$.

PROOF. Consider a function $g \in W_{\varepsilon\delta}([t_m, t_{m+n}])$ such that

$$c(g) = (0, ..., 0).$$

By Theorem 1.2 for the rth $r = 1, 2, ...$ definite integral of the function g at a point $t_{m+k}, k = 1, ..., n$ we get

$$g^{[r]}(t_{m+k}) = \sum_{i=1}^{r} \beta_{rki} g^{[i]}(t_m) \varepsilon^{r-i}. \tag{1.43}$$

On the other hand, it is easy to see that after r integrations of the function g, we have

$$g^{[r]}(t_{m+k}) = \sum_{i=1}^{r} k^{r-i} g^{[i]}(t_m) \varepsilon^{r-i}/(r-i)!. \tag{1.44}$$

Combining (1.43) and (1.44), we arrive at

$$\sum_{i=1}^{r} g^{[i]}(t_m)(\beta_{rki} - k^{r-i}/(r-i)!)\varepsilon^{r-i} = 0. \tag{1.45}$$

Recall that $g^{[i]}(t_m), i = 1, ..., r$ can be prescribed arbitrary. Consequently, for each $i = 1, ..., r$ we can set $g^{[i]}(t_m)$ to be nonzero and

if $i = 1,\ r \geq 2$ then

$$g^{[j]}(t_m) = 0, \quad j = i + 1, ..., r,$$

if $i = 2,\ r = 2$ then

$$g^{[j]}(t_m) = 0, \quad j = 1,$$

if $i \geq 2,\ r \geq 2$ then

$$g^{[j]}(t_m) = 0, \quad j = 1, ..., i - 1, i + 1, ..., r.$$

From this consideration, by using (1.45), we finally obtain

$$\beta_{rki} = \frac{k^{r-i}}{(r-i)!}, \quad i = 1, ..., r.$$

This completes the proof of the theorem.

3. HOW ICS GENERATES INTEGER RELATIONS

When graphs of functions have symmetry, sometimes it is literally possible to see values of definite integrals as they appear to be equal zero. The merit of the integer code series is that it translates this result into integer relations. In this section we illustrate this specific potential of ICS by showing how it can be used to make Pythagorean triples appear in a completely different context. This prepares us for the following chapters where integer relations generated by ICS become main objects of interest.

The illustrations are designed to show that integer relations are connected with two-dimensional geometrical patterns and their symmetry. Interestingly, well-known integer relations in these illustrations may be viewed to appear as broken grounds of the symmetry.

For convenience let $\varepsilon = 1$ and $\delta = 1$. Clearly then $t_k = k\varepsilon = k$, where k is an integer. Consider a function $g \in W_{11}([k_1, k_5])$ such that

$$g(t) = \begin{cases} 1, & t \in [k_1, k_1 + 1] \\ 0, & t \in (k_1, k_2 - 1] \\ -1, & t \in (k_2 - 1, k_3] \\ 1, & t \in (k_3, k_4 + 1] \\ 0, & t \in (k_4 + 1, k_5 - 1] \\ -1, & t \in (k_5 - 1, k_5] \end{cases}$$

and

$$g^{[r]}(k_1) = 0, \quad r = 1, 2, 3, \tag{1.46}$$

where $m = k_1$, $n = k_5 - k_1$. Integers $k_i, i = 1, ..., 5$, given in the definition of the function g, are specified by conditions

$$k_2 - k_1 = k_5 - k_4, \quad k_3 - k_2 = k_4 - k_3,$$

$$\int_{k_1}^{k_2} g^{[1]}(t)dt = - \int_{k_2}^{k_3} g^{[1]}(t)dt,$$

$$\int_{k_2}^{k_3} g^{[1]}(t)dt = \int_{k_3}^{k_4} g^{[1]}(t)dt = - \int_{k_4}^{k_5} g^{[1]}(t)dt. \tag{1.47}$$

Conditions (1.47) are set to ensure that

$$g^{[1]}(k_5) = 0,$$

$$g^{[2]}(k_5) = \int_{k_1}^{k_5} g^{[1]}(t)dt = 0,$$

$$g^{[3]}(k_5) = \int_{k_1}^{k_5} g^{[2]}(t)dt = 0. \tag{1.48}$$

From geometrical and symmetrical consideration it is easy to see that the specification is possible.

Evidently, from (1.47) for $k_i, i = 1, ..., 5$ we obtain

$$\int_{k_1}^{k_2} g^{[1]}(t)dt = \frac{1}{2} + (k_2 - k_1 - 2) + \frac{1}{2} = k_2 - k_1 - 1,$$

$$\int_{k_2}^{k_3} g^{[1]}(t)dt = \frac{(k_3 - k_2)^2}{2}.$$

By using (1.47), we get

$$k_2 - k_1 - 1 = \frac{(k_3 - k_2)^2}{2}. \tag{1.49}$$

Similarly, we have

$$k_5 - k_4 - 1 = \frac{(k_4 - k_3)^2}{2}.$$

Since $k_2 - k_1 - 1$ is an integer then due to (1.47) we obtain that $k_3 - k_2$ is an even number. Let

$$k_3 - k_2 = 2l, \quad l = 1, 2, ... \tag{1.50}$$

then

$$k_2 - k_1 - 1 = 2l^2, \quad l = 1, 2, ... \tag{1.51}$$

Analogously,

$$k_4 - k_3 = 2l, \quad k_5 - k_4 - 1 = 2l^2, \quad l = 1, 2, ... \tag{1.52}$$

Next, we express conditions (1.48) in terms of the integer code series. Let $c(g) = s_1...s_n$ be the code of the function g. Then by definition of the function g we have

$$s_{k_5+1-i} = 1, \quad i = k_1 + 1, k_3 + 1, \ k_3 + 2, ..., k_4 + 1,$$

$$s_{k_5+1-i} = -1, \quad i = k_2, k_2 + 1, ..., k_3 - 2, k_3 - 1, k_3, k_5,$$

$$s_{k_5+1-i} = 0, \quad i = k_1 + 1, ..., k_5,$$

$$i \neq k_1 + 1, \ k_2, k_2 + 1, ..., k_3, k_3 + 1, ..., k_4 + 1, \ k_5. \tag{1.53}$$

Theorem 2.1 for (1.48), by using (1.46), reads.

Firstly, for $g^{[1]}(k_5)$ we get

$$g^{[1]}(k_5) = \alpha_{1k_10}(k_5^0 s_1 + (k_5 - 1)^0 s_2 + \ldots + (k_1 + 1)^0 s_n) = 0.$$

Thus, since $\alpha_{1k_10} \neq 0$,

$$k_5^0 s_1 + (k_5 - 1)^0 s_2 + \ldots + (k_1 + 1)^0 s_n = 0. \tag{1.54}$$

Secondly, for $g^{[2]}(k_5)$ we find

$$\begin{aligned} &g^{[2]}(k_5) \\ &= \alpha_{2k_11}(k_5^1 s_1 + (k_5 - 1)^1 s_2 + \ldots + (k_1 + 1)^1 s_n) \\ &+\alpha_{2k_10}(k_5^0 s_1 + (k_5 - 1)^0 s_2 + \ldots + (k_1 + 1)^0 s_n) = 0. \end{aligned}$$

Hence, from (1.54) and since $\alpha_{2k_11} \neq 0$,

$$k_5^1 s_1 + (k_5 - 1)^1 s_2 + \ldots + (k_1 + 1)^1 s_n = 0. \tag{1.55}$$

Thirdly, for $g^{[3]}(k_5)$ we have

$$\begin{aligned} &g^{[3]}(k_5) \\ &= \alpha_{3k_12}(k_5^1 s_2 + (k_5 - 1)^2 s_2 + \ldots + (k_1 + 1)^2 s_n) \\ &+\alpha_{3k_11}(k_5^1 s_1 + (k_5 - 1)^1 s_2 + \ldots + (k_1 + 1)^1 s_n) \\ &+\alpha_{3k_10}(k_5^0 s_1 + (k_5 - 1)^0 s_2 + \ldots + (k_1 + 1)^0 s_n) = 0. \end{aligned}$$

Then, by using (1.54) and (1.55) with $\alpha_{3k_12} \neq 0$, we arrive at

$$k_5^2 s_1 + (k_5 - 1)^2 s_2 + \ldots + (k_1 + 1)^2 s_n = 0. \tag{1.56}$$

Combining (1.54), (1.55) and (1.56), we can write them as a system of relations between integers $k_1 + 1, k_1 + 2, \ldots, k_5$

$$\begin{aligned} s_1 k_5^0 + s_2(k_5 - 1)^0 + \ldots + s_n(k_1 + 1)^0 &= 0 \\ s_1 k_5^1 + s_2(k_5 - 1)^1 + \ldots + s_n(k_1 + 1)^1 &= 0 \\ s_1 k_5^2 + s_2(k_5 - 1)^2 + \ldots + s_n(k_1 + 1)^2 &= 0 \end{aligned} \tag{1.57}$$

On the one hand, conditions (1.48), expressed by using ICS in terms of (1.57), are due to symmetry imposed on the graph of the first integral of the function g. Upon two integrations of the first integral this symmetry successively evolves into symmetries that graphs of the second and third integrals of the function have.

Therefore, it can be said that ICS translates these symmetries into the system of integer relations (1.57). On the other hand, looking at the system (1.57) we can feel its harmony. But it is a different matter to put this harmony into pictures that can express it directly. ICS provides such an opportunity for the system by connecting it with symmetries of the graphs.

The system (1.57) viewed by itself does not exhibit the symmetries. Moreover, rules of arithmetic are flexible to make the situation even worse. We explain what this means by focusing on the last relation of (1.57).

Regrouping it and taking into account (1.53), we obtain

$$(k_1+1)^2+(k_3+1)^2+(k_3+2)^2+...+(k_4+1)^2$$

$$=k_3^2+(k_3-1)^2+(k_3-2)^2+...+(k_2+1)^2+k_2^2+k_5^2. \qquad (1.58)$$

Notice that these manipulations preserve in (1.58) the harmony of (1.56). Let us specify (1.58) by $k_3=0$. This, by using (1.47), gives $k_4=-k_2$. Then for (1.58) we have

$$(k_1+1)^2+1^2+2^2+...+(k_4+1)^2$$

$$=0^2+(-1)^2+(-2)^2+...+(k_2+1)^2+k_2^2+k_5^2. \qquad (1.59)$$

The integer relation (1.59) is a special case of (1.58) and still has its form.

Now, we make a step that is usually done in simplifying algebraic expressions. This step in our case has a new meaning. Observe that in (1.59), due to rules of arithmetic, we can exclude term by term integers

$$1^2, 2^2, ..., (k_4+1)^2$$

from the left side simultaneously with integers

$$(-1)^2, (-2)^2, ..., (-(k_4+1))^2$$

from the right side. These exclusions are possible because they give a sequence of true relations resulting in

$$(k_1+1)^2+(k_4+1)^2=0^2+k_5^2. \qquad (1.60)$$

Usually we do not keep integer 0 as a term in algebraic expressions. This breaks harmony in (1.60) as the number of terms in the left side and does not equal the number of terms in the right side and we obtain

$$(k_1 + 1)^2 + (k_4 + 1)^2 = k_5^2. \tag{1.61}$$

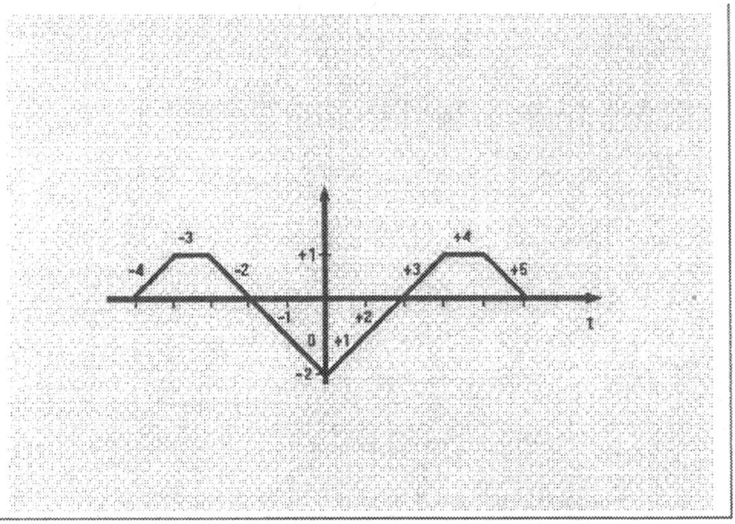

Figure 1: Graph of the first integral of the function g for $l = 1$.

In the following chapters we show that integer relations can be perceived as geometrical patterns with a specific rule of composition. In such an integer relation each integer and its powers matter as they are integral parts of a process in which the integer relation and a corresponding geometrical pattern are formed. Algebraicly, relations (1.59) and (1.61) are the same. But although the first one can be seen as a geometrical object the second one cannot be.

By using (1.50), (1.51) and (1.52), we get

$$k_1 = -1 - 2l - 2l^2, \quad k_4 = 2l, \quad k_5 = 1 + 2l + 2l^2,$$

where $l = 1, 2, \ldots$.

Within the standard notation for Pythagorean triples we have

$$x = k_4 + 1 = 2l + 1, \quad -y = k_1 + 1 = 2l + 2l^2,$$

$$z = k_5 = 1 + 2l + 2l^2, \quad l = 1, 2, \ldots .$$

Thus we obtain a subset of Pythagorean triples

$$x = k_4 + 1 = 2l + 1, \quad y = k_1 + 1 = 2l + 2l^2,$$

$$z = k_5 = 1 + 2l + 2l^2, \quad l = 1, 2, \ldots . \tag{1.62}$$

On Figures 1 and 2 the function $g^{[1]}$ is depicted for $l = 1, 2$.

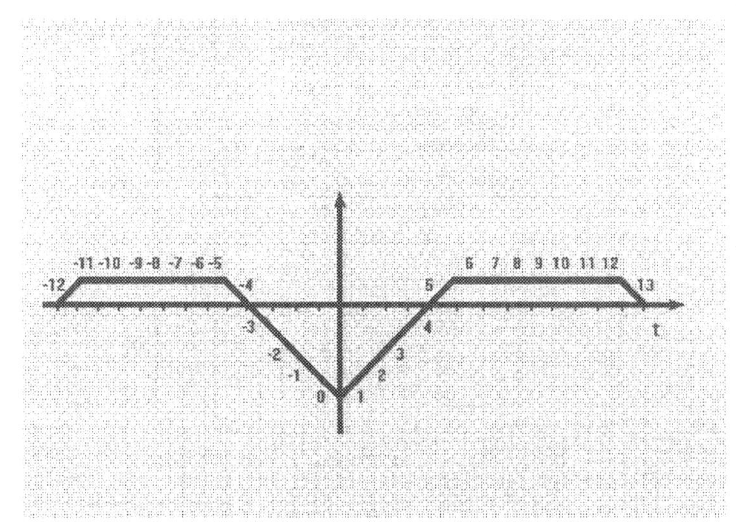

Figure 2: Graph of the first integral of the function g for $l = 2$.

Finally, given a subset of Pythagorean triples, we illustrate how ICS can generate another subset of Pythagorean triples. This illustration gives a picture of two-dimensional geometrical patterns and is helpful to the next chapter where the geometrical patterns are defined formally.

Consider a function $g_1 \in W_{11}([k_1, k_6])$ such that

$$g_1(t) = \begin{cases} 1, \ t \in [k_i, k_i + 1], i = 1, 5, \\ 0, \ t \in (k_i + 1, k_{i+1} - 1], i = 1, 5, \\ -1, \ t \in (k_i - 1, k_i], i = 2, 6, \\ 0, \ t \in (k_i, k_{i+1}], i = 2, 4, \\ -1, \ t \in (k_3, k_3 + 1] \\ 0, \ t \in (k_3 + 1, k_4 - 1] \\ 1, \ t \in (k_4 - 1, k_4] \end{cases}$$

and

$$g_1^{[r]}(k_1) = 0, \quad r = 1, 2, 3,$$

where $m = k_1$, $n = k_6 - k_1$ and $k_i, i = 1, ..., 6$ are integers.

The first integral of the function g_1 looks like a symmetrical collection of two-dimensional geometrical patterns and is given by (see Figure 3)

$$g_1^{[1]}(t) = \begin{cases} t - k_i, \ t \in [k_i, k_i + 1], i = 1, 5, \\ 1, \ t \in (k_i + 1, k_{i+1} - 1], i = 1, 5, \\ -t + k_i, \ t \in (k_i - 1, k_i], i = 2, 6, \\ 0, \ t \in (k_i, k_{i+1}], i = 2, 4, \\ -t + k_3, \ t \in (k_3, k_3 + 1] \\ -1, \ t \in (k_3 + 1, k_4 - 1] \\ t - k_4, \ t \in (k_4 - 1, k_4] \end{cases}$$

In the definition of the function g_1 integers $k_i, i = 1, ..., 6$ are chosen

$$k_2 - k_1 = k_6 - k_5, \quad k_3 - k_2 = k_5 - k_4,$$

$$k_4 - k_3 = k_2 - k_1 + k_6 - k_5 - 1, \tag{1.63}$$

to specify the following conditions

$$g_1^{[r]}(k_6) = 0, \quad r = 1, 2, 3. \tag{1.64}$$

From geometrical and symmetrical consideration it is easy to see that the specification is possible.

Let $c(g_1) = s_1 ... s_n$ be the code of the function g_1. By definition of the function g_1 we have

$$s_{k_6+1-i} = 1, \quad i = k_1 + 1, k_4, k_5 + 1,$$

$$s_{k_6+1-i} = -1, \quad i = k_2, k_3 + 1, k_6,$$

$$s_{k_6+1-i} = 0, \quad i = k_1 + 1, ..., k_6,$$

$$i \neq k_1 + 1, k_2, k_3 + 1, k_4, k_5 + 1, k_6. \tag{1.65}$$

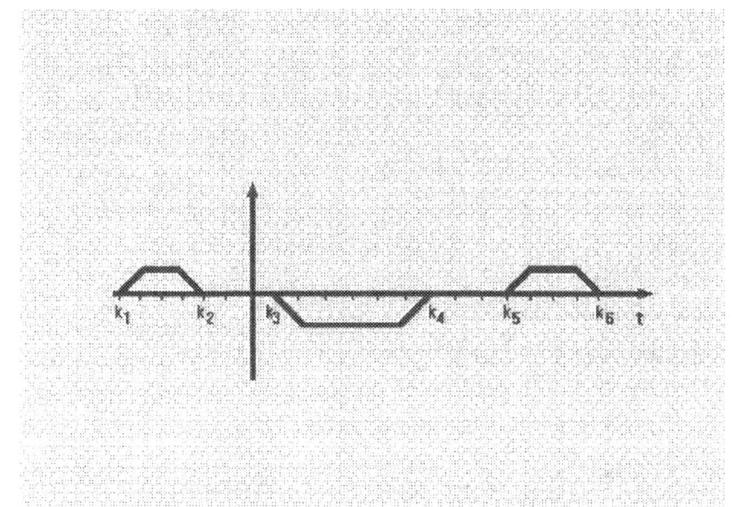

Figure 3: Graph of the first integral of the function g_1.

Expressing conditions (1.64) in terms of ICS and applying the same recipe as for the function g in the previous illustration, similarly to (1.56), we obtain

$$k_6^2 s_1 + (k_6 - 1)^2 s_2 + \dots + (k_1 + 1)^2 s_n = 0. \tag{1.66}$$

In view of (1.65), rearranging (1.66), we come to

$$(k_1 + 1)^2 + k_4^2 + (k_5 + 1)^2 = k_2^2 + (k_3 + 1)^2 + k_6^2. \tag{1.67}$$

For standard appearance let

$$z_2 = k_1 + 1, \quad x_1 = k_4, \quad y_1 = k_5 + 1,$$

$$z_1 = k_2, \quad x_2 = k_3 + 1, \quad y_2 = k_6. \tag{1.68}$$

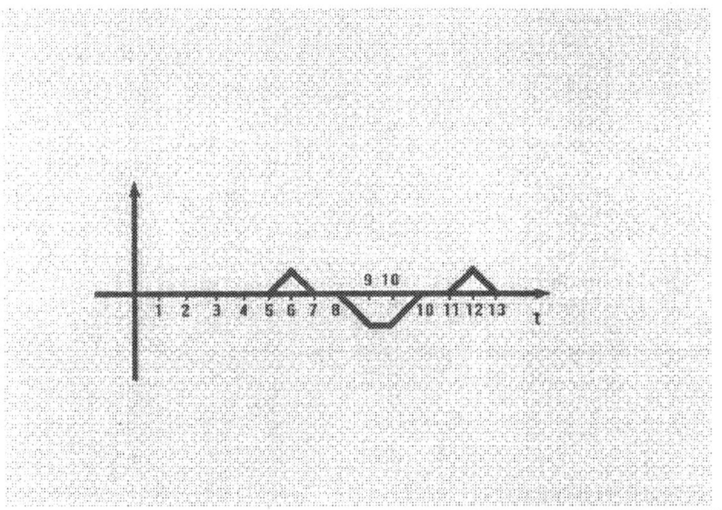

Figure 4: Graph of the first integral of the function g_1 for $l = 2$.

By using (1.68), we write (1.67) as

$$x_1^2 + x_2^2 + x_3^2 = y_1^2 + y_2^2 + y_3^2. \tag{1.69}$$

Assume, for example, that the triple (y_2, y_3, x_1) is Pythagorean, namely

$$y_2^2 + y_3^2 = x_1^2. \tag{1.70}$$

This initiates symmetry breaking of (1.69) because by using (1.70) we can simply write (1.69) as

$$x_2^2 + x_3^2 = y_1^2.$$

Thus, the triple (x_2, x_3, y_1) is also Pythagorean. Consequently, if we take (y_2, y_3, x_1) from the subset of Pythagorean triples defined by (1.62), then we get another subset of Pythagorean triples.

By direct computation it is easy to see, using (1.63) and (1.68), that this subset reads

$$x = 2l + 2, \quad y = 2l + 2l^2,$$
$$z = 2 + 2l + 2l^2, \quad l = 2, 3, \ldots .$$

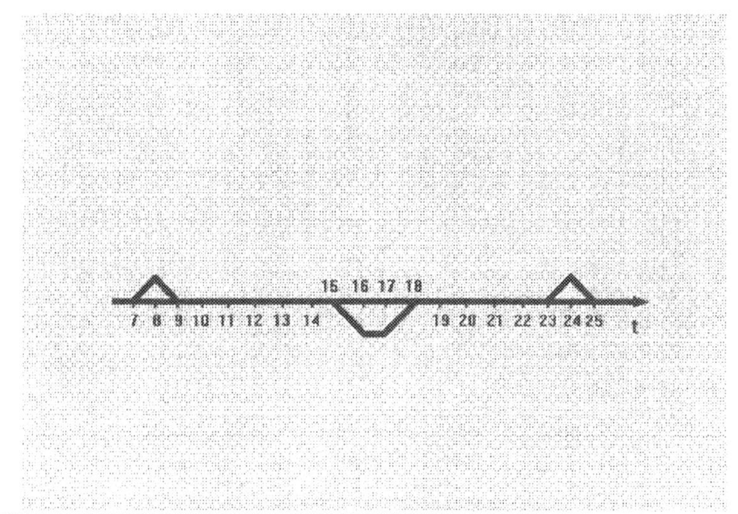

Figure 5: Graph of the first integral of the function g_1 for $l = 2$.

On Figures 4 and 5 the function $g_1^{[1]}$ is depicted for $l = 2, 3$. As l changes the central pattern stays as the left and right one moves correspondingly to preserve symmetry of the whole arrangement of the patterns.

Chapter 2

SYSTEMS OF INTEGER RELATIONS AND STRUCTURAL COMPLEXITY

The notion of sequence naturally arises from Cartesian order and plays a very important role in the study of natural systems. In this chapter the integer code series is used to propose a global description of sequences as the next step towards the definition of the structure. Namely, given in section 1 this global description of sequences produces in section 2 systems of integer relations. These systems of integer relations become the main ingredient in the definition of the structure.

The key idea concerning the systems is that they represent a new type of hierarchical formation of integer relations and two-dimensional geometrical patterns. In the chapter first steps are made to consider these hierarchical formations. As the character of the hierarchical formations turns out to be very close to what is associated with complexity, they are proposed to measure the complexity of natural systems. For this purpose a concept of complexity, called structural complexity, is introduced. The concept, presented and illustrated in section 3, is useful in describing the hierarchical formations and plays a very important role.

In section 4 it is shown that the Prouhet-Thue-Morse (PTM) sequence has a remarkable property in terms of structural complexity. By using period doubling the property connects structural complexity with self-organization and supports the concept to be used as a measure of complexity.

In section 5 two-dimensional geometrical patterns, called integer patterns, unfolding the systems of integer relations into visible images are presented. This prepares us to perceive integer relations as geometrical objects that can form into each other.

In section 6 hierarchical formations for a particular system of integer relations are visualized. This gives an interesting picture containing nonlocal order and large symmetry. The picture hopefully gives us insight into the harmony of the structure before its definition.

—

1. STRUCTURAL NUMBERS AS A GLOBAL DESCRIPTION OF SEQUENCES

Sequences usually occur when natural systems are analysed in terms of space-time as independent and interacting by forces parts. This kind of analysis is often associated with Cartesian order (for example, [8]). The structure presented in this book is proposed to analyse natural systems not as independent but as integrated parts of a unified whole.

In this section we start to set up a connection between a description of a natural system in terms of space-time with a description of the system in the structure. There is a sequence in space-time on the one hand and there are hierarchical formations of integer relations and geometrical patterns of the structure on the other hand.

Let I be an integer alphabet and

$$I_n = \{s : s = s_1...s_n, s_i \in I, i = 1, ..., n\}$$

be the set of all sequences of length $n \geq 1$. If $I = \{-1, +1\}$ then I_n is the set of all binary sequences of length n, specially denoted B_n. Let $s(i) = s_1...s_i$ and $|s(i)| = i$ denote the length of $s(i) \in I_i, i = 1, ..., n$. For convenience for all $s \in I_n$ let $s_0 = \emptyset$, where $\emptyset$ is the empty sequence.

A sequence gives a simple description concerning the dynamics of a natural systems and is used as an approximation to its trajectory in space-time. Namely, let δ and ε be respective spacings of a space-time lattice (δ, ε) in $1+1$ dimensions. A sequence $s = s_1...s_n \in I_n$ encodes on the lattice (δ, ε) the dynamics of a natural system that has a space position $s_i\delta$ at a time instant $i\varepsilon, i = 1, ..., n$.

A global description of sequences is a representation that uses information about sequences as whole objects. A *global description of sequences* proposed in [14] and considered in the book associates a sequence

$$s = s_1...s_n \in I_n$$

with numbers

$$\vartheta_1(s), ..., \vartheta_k(s), ... ,$$

called *"structural" numbers*

$$s = s_1...s_n \Longrightarrow \vartheta_1(s), ..., \vartheta_k(s), ... ,$$

which capture global properties of the sequence on a lattice (δ, ε).

It will be shown that the numbers indeed play a very important role. This explains why we dare to use the ambitious term "structural" for the numbers. The definition of structural numbers is based on the integer code series and on the natural representation of sequences considered in Chapter 1.

Let a mapping $\rho_{m\delta\varepsilon}$ take a sequence

$$s = s_1...s_n, \quad s_i \in \Re^1, \quad i = 1, ..., n$$

to a function $W_{\delta\varepsilon}[t_m, t_{m+n}]$, denoted $f = \rho_{m\delta\varepsilon}(s)$, such that $c(f) = s$ and whose kth integral satisfies (see Figure 6)

$$f^{[k]}(t_m) = 0, \quad k = 1, 2, ...$$

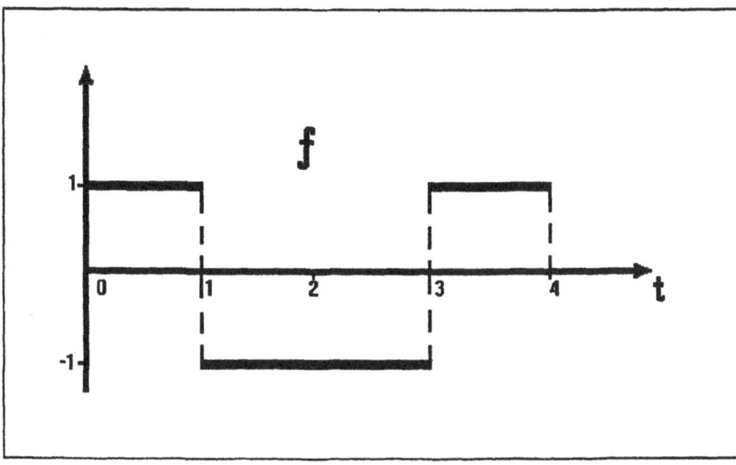

Figure 6: Graph of a function $f = \rho_{011}(s)$, $s = c(f) = s_1...s_4 = +1 - 1 - 1 + 1$, $\delta = 1$, $\varepsilon = 1, m = 0$, $n = 4$, $t_{m+n} = 4$.

Definition 2.1 (Structural Numbers). *Structural numbers*

$$\vartheta_k(s), \quad k = 1, 2, ...$$

of a sequence $s = s_1...s_n \in I_n$ *on a lattice* (δ, ε) *are defined by the formula*

$$\vartheta_k(s) = \sum_{i=0}^{k-1} \alpha_{kmi}((m+n)^i s_1 + (m+n-1)^i s_2 + ... + (m+1)^i s_n)\varepsilon^k \delta. \quad (2.1)$$

It is easy to see from the definition that there exists a

Geometrical Interpretation of Structural Numbers. *The kth structural number* $\vartheta_k(s), k = 1, 2, ...$ *of a sequence* $s \in I_n$ *on a lattice* (δ, ε) *is the value of the* kth *definite integral of a function*

$$f = \rho_{m\delta\varepsilon}(s) \in W_{\delta\varepsilon}[t_m, t_{m+n}]$$

at the point t_{m+n} [14]

$$\vartheta_k(s) = f^{[k]}(t_{m+n}), \quad k = 1, 2, ... \quad (2.2)$$

To illustrate the geometrical interpretation, consider the function f in Figure 6. The symmetry of the function is easily seen and we can sketch, thinking in geometrical images, the graph of its first $f^{[1]}$ and second $f^{[2]}$ integral as shown in Figures 7 and 8. From the figures we conclude that

$$f^{[1]}(t_4) = 0, \quad f^{[2]}(t_4) = 0, \quad f^{[3]}(t_4) > 0$$

and, by using (2.2),

$$\vartheta_1(s) = 0, \quad \vartheta_2(s) = 0, \quad \vartheta_3(s) > 0. \quad (2.3)$$

Note that the visualization gives much more than formula (2.1) by blind computation. For example, from Figures 6, 7 and 8 condition (2.3) is immediate to our intuition.

Actually, Definition 2.1 is chosen to ensure the validity of (2.2). Structural numbers can be seen as global properties of the sequence, since as definite integrals, they carry information about it as a whole. Due to Theorem 1.2 they are independent of integer m, which specifies a "frame of reference" for their consideration. *The geometrical interpretation of structural numbers is of crucial importance to the study.*

It seems surprising that we use definite integrals to describe global properties of sequences and even propose a special name for them. Definite integral is a well-known notion and is not expected very much to produce something new. However, it turns out that the notion of definite integral plays a more important role then we are used to. In particular, it combines together geometry and integers. Firstly, the definite integral of a piecewise constant function has the visualisable area interpretation and secondly due to Theorem 1.2 it can be expressed in terms of integers.

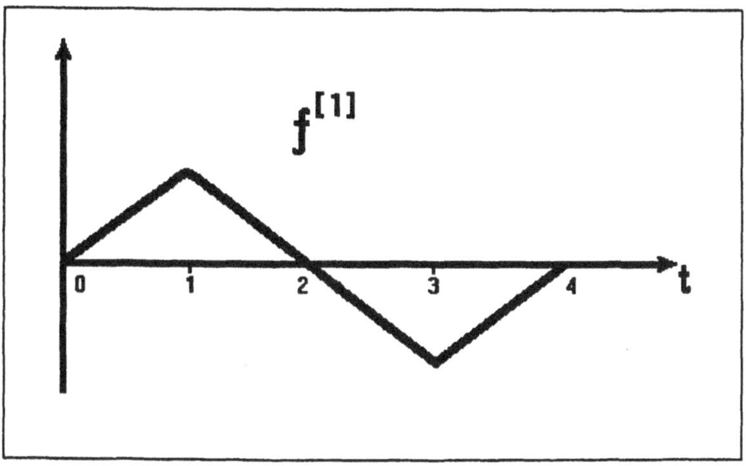

Figure 7: Graph of the first $f^{[1]}$ integral of the function f depicted in Figure 6.

It is helpful for us to place structural numbers in familiar contexts. Now, we consider such a context. Let

$$H(s) = \#\{j : s_j \neq 0,\ j = 1, ..., n\}.$$

For a sequence $s = s_1...s_n \in I_n$ let $\tau(s, j)$ be equal i such that s_i is the jth component of the sequence that does not equal zero as index i takes successive values $1, ..., n$. Let $\tau(s, 0) = 0$.

Assume for a sequence $s = s_1...s_n \in I_n$ we have $s_i \geq 0, i = 1, ..., n$ and $H(s) > 0$. Then (2.1) can be transformed into the following form

$$\vartheta_k(s) = \sum_{i=0}^{k-1} \alpha_{kmi}(x_1^i + ... + x_N^i)\varepsilon^k\delta, \tag{2.4}$$

where for $j = 1, ..., H(s)$

$$x_i = (m + n - \tau(s, j) + 1) \tag{2.5}$$

when

$$1 + \sum_{l=0}^{j-1} s_{\tau(s,l)} \leq i \leq \sum_{l=0}^{j} s_{\tau(s,l)}, \quad N = \sum_{j=1}^{H(s)} s_{\tau(s,j)}.$$

By using (2.5), the sequence s can be associated with the ith $i = 0, 1, ...$ power-sum of $x_1, ..., x_N$

$$P_i(s) = x_1^i + ... + x_N^i. \tag{2.6}$$

Inserting (2.6) into (2.4) gives

$$\vartheta_k(s) = \sum_{i=0}^{k-1} \alpha_{kmi} P_i(s)\varepsilon^k\delta. \tag{2.7}$$

From (2.7) we see that if power sums

$$P_0(s), ..., P_{k-1}(s), \quad k = 1, 2, ...$$

of the sequence s are known then its structural numbers $\vartheta_1(s), ..., \vartheta_k(s)$ are known also. It is easy to prove by induction on $i = 1, ..., k$ that the inverse statement is also true.

For $i = 1$ from (2.7) we have immediately

$$P_0(s) = \frac{\vartheta_1(s)}{\alpha_{1m0}\varepsilon\delta}.$$

If $k = 1$ we are done, otherwise assume that for $1 \leq i < k$ structural numbers $\vartheta_1(s), ..., \vartheta_i(s)$ specify power sums $P_0(s), ..., P_{i-1}(s)$, i.e., there are functions $G_0, ..., G_{i-1}$ such that

$$P_0(s) = G_0(\vartheta_1(s)), ..., P_{i-1}(s) = G_{i-1}(\vartheta_1(s), ..., \vartheta_i(s)). \tag{2.8}$$

For $i + 1$ from (2.7) and the inductive assumption (2.8) we get that the ith power sum

$$P_i(s)$$

$$= \frac{1}{\alpha_{i+1,m_i}\varepsilon^{i+1}\delta}(\vartheta_{i+1}(s) - \sum_{j=0}^{i-1} \alpha_{i+1,m_j} P_j(s)\varepsilon^{i+1}\delta)$$

$$= \frac{1}{\alpha_{i+1,m_i}\varepsilon^{i+1}\delta}(\vartheta_{i+1}(s) - \sum_{j=0}^{i-1} \alpha_{i+1,m_j} G_j(\vartheta_1(s), ..., \vartheta_{j+1}(s))$$

$$= G_i(\vartheta_1(s), ..., \vartheta_{i+1}(s))$$

is specified by structural numbers $\vartheta_1(s), ..., \vartheta_{i+1}(s)$, as required.

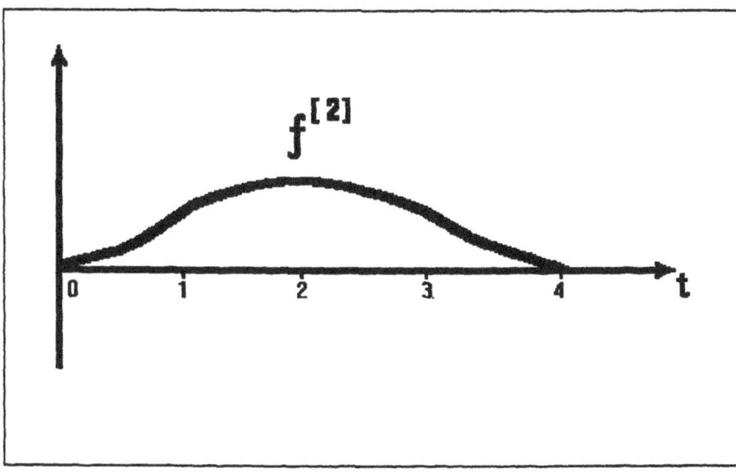

Figure 8: Graph of the second $f^{[2]}$ integral of the function f depicted in Figure 6.

Thus for each $k = 1, 2, ...$ we have a one-to-one correspondence between structural numbers of a sequence $s = s_1...s_n \in I_n$ and its power sums (2.6)

$$\{\vartheta_1(s), ..., \vartheta_k(s)\} \iff \{P_0(s), ..., P_{k-1}(s)\}, \tag{2.9}$$

where $s_i \geq 0, i = 1, ..., n$ and $H(s) > 0$.

Consider a polynomial with integer roots $x_1, ..., x_N$, so that

$$x^N + \sigma_1 x^{N-1} + ... + \sigma_N = (x - x_1)...(x - x_N). \tag{2.10}$$

It is well-known that the coefficients $\sigma_1, ..., \sigma_N$ are the polynomial functions in $x_1, ..., x_N$, called the elementary symmetric polynomials [32],

$$\sigma_i = (-1)^i \sum_{1 \le j_1 < j_2 ... < j_i \le N} x_{j_1} x_{j_2} ... x_{j_i}, \quad i = 1, ..., N.$$

The power-sum

$$P_i = x_1^i + ... + x_N^i, \quad i = 1, 2, ...$$

is clearly a symmetric polynomial and it can be expressed by the fundamental theorem on symmetric polynomials as a unique polynomial in the elementary symmetric polynomials. Moreover, the elementary symmetric polynomials and the power-sums are connected by Newton's identities, which provide a recursive scheme for expressing P_i in terms of $\sigma_i, i = 1, ..., N$ and vice versa [32]. Thus

$$\{\sigma_1, ..., \sigma_N\} \Longleftrightarrow \{P_1, ..., P_N\}. \tag{2.11}$$

It is possible to encode information about the roots of the polynomial (2.10) with the help of a piecewise constant function. Let $\varepsilon = 1, \delta = 1$ and consider an interval $[t_m, t_{m+n}]$, where

$$n = \max_{i=1,...,N} x_i - \min_{i=1,...,N} x_i + 1 \ge 2,$$

$$m = \min_{i=1,...,N} x_i - 1,$$

then a function f, called spectral,

$$f : [t_m, t_{m+n}] \to \Re^1$$

that encodes the roots is defined in an interval

$$[t_{2m+n-k}, t_{2m+n-k+1}),$$

$$k = \max_{i=1,...,N} x_i, \max_{i=1,...,N} x_i - 1, ..., \min_{i=1,...,N} x_i$$

as the number of the roots that are equal to k. The value of the function at the point t_{m+n} is equal to its value on the interval $[t_{m+n-1}, t_{m+n})$.

For example, in Figure 9 a spectral function that encodes roots

$$x_1 = -2, \quad x_2 = -1, \quad x_3 = -1,$$

$$x_4 = -1, \quad x_5 = +1, \quad x_6 = +4$$

of a polynomial is shown.

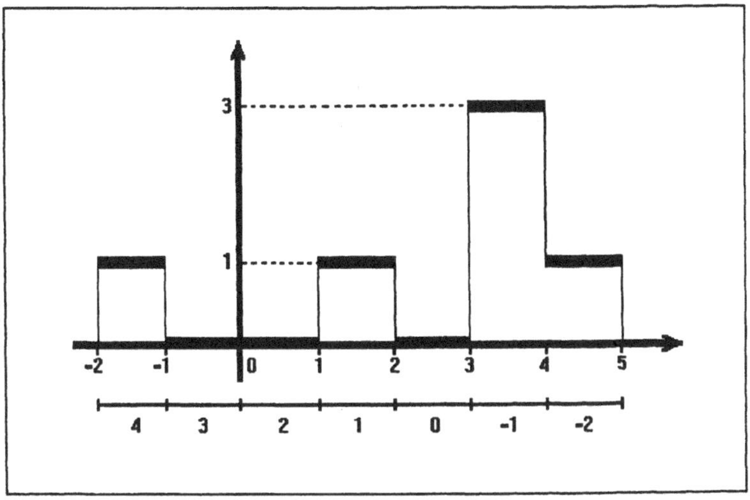

Figure 9: Spectral function encodes information about integer roots $(-2,-1,-1,-1,+1,+4)$ of a polynomial.

It follows from the definition that the spectral function f belong to $W_{11}[t_m, t_{m+n}])$. Let a sequence $s = s_1...s_n \in I_n$ be the code of a spectral function f. Then for structural numbers of the sequence s we have

$$\vartheta_k(s) = \sum_{i=0}^{k-1} \alpha_{kmi}((m+n)^i s_1 + ... + (m+1)^i s_n), \quad k = 1, 2, ... \quad (2.12)$$

where s_i is the number of the roots that are equal to

$$(m+n-i+1), \quad i = 1, ..., n.$$

This implies that (2.12) can be rearranged to give

$$\vartheta_k(s) = \sum_{i=0}^{k-1} \alpha_{kmi}(x_1^i + ... + x_N^i) = \sum_{i=0}^{k-1} \alpha_{kmi} P_i(s), \quad k = 1, 2, ...$$

Therefore, by taking into account (2.9), (2.11) and dependence of σ_i, $i = 1, ..., N$ on the sequence s, we have

$$\{\sigma_0(s), ..., \sigma_N(s)\} \iff \{P_0(s), ..., P_N(s)\}$$
$$\iff \{\vartheta_1(s), ..., \vartheta_{N+1}(s)\}$$

or

$$\{\sigma_0(s), ..., \sigma_N(s)\} \iff \{\vartheta_1(s), ..., \vartheta_{N+1}(s)\}$$

where $\sigma_0 = P_0(s) = N$.

This gives the following context for structural numbers.

The coefficients of a polynomial of degree N with integer roots are in a one-to-one correspondence with structural numbers of a sequence that encodes the roots of the polynomial.

It may be worth mentioning that Majorana proposed to consider the unordered set of N points on the Riemann sphere as the N roots of a complex polynomial of degree N. The coefficients of this polynomial are used then as coordinates of the $(N + 1)$-dimensional Hilbert space [33]. We have a simpler situation as the set of N points on the real line are considered as N integer roots of a polynomial.

2. STRUCTURAL NUMBERS AS A BASIS AND SYSTEMS OF INTEGER RELATIONS

In section 1 we presented structural numbers, which provided a global description for the sequence. We mentioned that structural numbers of a sequence, by a connection with the coefficients of a polynomial, may be seen as coordinates of a Hilbert space.

It turns out that the development of the structure further leads us to consider structural numbers as basis in a space. Namely, the question discussed in this section is: Is it possible to completely specify a sequence by a finite number of structural numbers? By answering the question we come to systems of integer relations that become the main ingredient in the definition of the structure. The following definition is useful.

Definition 2.2 *If for a pair of different sequences $s, s' \in I_n$ there is a finite integer $k \geq 2$ such that their first $(k - 1)$ structural numbers are equal*

$$\vartheta_1(s) = \vartheta_1(s'), ..., \vartheta_{k-1}(s) = \vartheta_{k-1}(s'),$$

whereas the kth *structural numbers are not*

$$\vartheta_k(s) \neq \vartheta_k(s')$$

then

$$C(s, s', n) = k.$$

If $\vartheta_1(s) \neq \vartheta_1(s')$ *then*

$$C(s, s', n) = 1.$$

In particular, Definition 2.2 implies that by using first $C(s, s', n)$ of structural numbers a sequence $s \in I_n$ can be completely specified with respect to another sequence $s' \in I_n$. Note that a special order is used in this specification of sequences as structural numbers are considered successively starting with the first ones. The following theorem clarifies the question about the ability of structural numbers to completely specify a sequence.

Theorem 2.1 (Structural Numbers as Basis). *Let*

$$s = s_1...s_n, \ s' = s'_1...s'_n \in I_n$$

be a pair of different sequences. Then

$$C(s, s', n) \leq n.$$

PROOF. Consider a sequence $s = s_1...s_n \in I_n$ and let for $i = 0, 1, ...$

$$\aleph_i(s, m) = (m+n)^i s_1 + (m+n-1)^i s_2 + ... + (m+1)^i s_n. \tag{2.13}$$

Then (2.1) for $k = 1, 2, ...$ takes the form

$$\vartheta_k(s) = \sum_{i=0}^{k-1} \alpha_{kmi} \aleph_i(s, m) \varepsilon^k \delta. \tag{2.14}$$

We first claim that structural numbers

$$\vartheta_i(s), \quad i = 1, ..., k, \quad k \geq 1$$

define numbers

$$\aleph_{i-1}(s, m), \quad i = 1, ..., k$$

and

for $k = 1$ if

$$\vartheta_1(s) = \vartheta_1(s') \tag{2.15}$$

then

$$\aleph_0(s, m) = \aleph_0(s', m), \tag{2.16}$$

for $k \geq 2$ if

$$\vartheta_1(s) = \vartheta_1(s'), ..., \vartheta_{k-1}(s) = \vartheta_{k-1}(s'), \tag{2.17}$$

$$\vartheta_k(s) \neq \vartheta_k(s'), \tag{2.18}$$

then

$$\aleph_0(s, m) = \aleph_0(s', m), ..., \aleph_{k-2}(s, m) = \aleph_{k-2}(s', m), \tag{2.19}$$

$$\aleph_{k-1}(s, m) \neq \aleph_{k-1}(s', m), \tag{2.20}$$

and vice versa.

We proceed with the claim by induction on $i = 1, ..., k$. The claim immediately follows for $i = 1$ from (2.14) and (2.15)

$$\aleph_0(s, m) = \frac{\vartheta_1(s)}{\alpha_{1m0}\varepsilon\delta} = \frac{\vartheta_1(s')}{\alpha_{1m0}\varepsilon\delta} = \aleph_0(s', m).$$

If $k = 1$ we are done, otherwise assume that for $1 \leq i < k$ structural numbers

$$\vartheta_1(s), ..., \vartheta_i(s)$$

define numbers

$$\aleph_0(s, m), ..., \aleph_{i-1}(s, m)$$

and if

$$\vartheta_1(s) = \vartheta_1(s'), ..., \vartheta_i(s) = \vartheta_i(s')$$

then

$$\aleph_0(s, m) = \aleph_0(s', m), ..., \aleph_{i-1}(s, m) = \aleph_{i-1}(s', m).$$

We prove it for $i + 1$. From (2.14) and the inductive assumption we immediately get that the number

$$\aleph_i(s, m) = \frac{1}{\alpha_{i+1,mi}\varepsilon^{i+1}\delta}(\vartheta_{i+1}(s) - \sum_{j=0}^{i-1} \alpha_{i+1,mj}\aleph_j(s, m)\varepsilon^{i+1}\delta)$$

is defined.

If $i+1<k$ then

$$\aleph_i(s,m)$$

$$= \frac{1}{\alpha_{i+1,mi}\varepsilon^{i+1}\delta}(\vartheta_{i+1}(s) - \sum_{j=0}^{i-1} \alpha_{i+1,mj}\aleph_j(s,m)\varepsilon^{i+1}\delta)$$

$$= \frac{1}{\alpha_{i+1,mi}\varepsilon^{i+1}\delta}(\vartheta_{i+1}(s') - \sum_{j=0}^{i-1} \alpha_{i+1,mj}\aleph_j(s',m)\varepsilon^{i+1}\delta)$$

$$= \aleph_i(s',m),$$

as required.

If $i+1=k$ then, in order to prove (2.20), we use (2.14) and (2.18)

$$\aleph_{k-1}(s,m) - \aleph_{k-1}(s',m)$$

$$= \frac{1}{\alpha_{km,k-1}\varepsilon^k\delta}(\vartheta_k(s) - \sum_{j=0}^{k-2} \alpha_{kmj}\aleph_j(s,m)\varepsilon^k\delta)$$

$$- \frac{1}{\alpha_{km,k-1}\varepsilon^k\delta}(\vartheta_k(s') - \sum_{j=0}^{k-2} \alpha_{kmj}\aleph_j(s',m)\varepsilon^k\delta)$$

$$= \frac{1}{\alpha_{km,k-1}}(\vartheta_k(s) - \vartheta_k(s')) \neq 0.$$

Conversely, for $k \geq 1$ we see from (2.14) that numbers

$$\aleph_{i-1}(s,m), \quad i=1,...,k$$

define structural numbers

$$\vartheta_i(s), \quad i=1,...,k$$

and (2.15) easily follows from (2.16) for $k=1$

$$\vartheta_1(s) = \alpha_{1m0}\aleph_0(s,m)\varepsilon\delta = \alpha_{1m0}\aleph_0(s',m)\varepsilon\delta = \vartheta_1(s').$$

Next, we prove (2.17) and (2.18) given (2.19) and (2.20). Indeed, for $k \geq 2$, since (2.19), we obtain (2.17)

$$\vartheta_i(s) = \sum_{j=0}^{i-1} \alpha_{imj}\aleph_j(s,m)\varepsilon^i\delta = \sum_{j=0}^{i-1} \alpha_{imj}\aleph_j(s',m)\varepsilon^i\delta = \vartheta_i(s'),$$

where $i=1,...,k-1$. We easily get

$$\vartheta_k(s) - \vartheta_k(s')$$

$$= \alpha_{km,k-1}\aleph_{k-1}(s,m)\varepsilon^k\delta + \sum_{i=0}^{k-2}\alpha_{kmi}\aleph_i(s,m)\varepsilon^k\delta$$

$$-\alpha_{km,k-1}\aleph_{k-1}(s',m)\varepsilon^k\delta - \sum_{i=0}^{k-2}\alpha_{kmi}\aleph_i(s',m)\varepsilon^k\delta,$$

which proves (2.18), by taking care of (2.19) and (2.20)

$$\vartheta_k(s) - \vartheta_k(s') = \alpha_{km,k-1}(\aleph_{k-1}(s,m) - \aleph_{k-1}(s',m))\varepsilon^k\delta \neq 0.$$

This finishes the claim.

By subtracting $\aleph_i(s',m), i = 0, ..., k-1$ from both sides of (2.19), (2.20) and using (2.13), we get a system of $(k-1)$ linear equations

$$(m+n)^i(s_1 - s_1') + ... + (m+1)^i(s_n - s_n') = 0, \tag{2.21}$$

where $i = 0, ..., k-2$ and an inequality

$$(m+n)^{k-1}(s_1 - s_1') + ... + (m+1)^{k-1}(s_n - s_n') \neq 0 \tag{2.22}$$

in n integer unknowns $s_i - s_i', i = 1, ..., n$.

Note that if the system (2.21) has n equations then at least one of its equations is nonzero. Indeed, the determinant of the Vandermonde matrix

$$\mathrm{V} = \begin{pmatrix} (m+1)^0 & (m+2)^0 & ... & (m+n)^0 \\ (m+1)^1 & (m+2)^1 & ... & (m+n)^1 \\ ... & ... & ... & ... \\ (m+1)^{n-1} & (m+2)^{n-1} & ... & (m+n)^{n-1} \end{pmatrix}$$

is known to be written in the form

$$det(V) = \pm\Pi\{i - j : 1 \leq j < i \leq n\} \neq 0 \tag{2.23}$$

and is nonzero [34]. Suppose all equations of the linear system are zero, then we conclude, with the help of (2.23), that the unique solution to the system is the trivial one $s_i - s_i' = 0, i = 1, ..., n$.

This contradicts that sequences s, s' are different. Thus not all equations of the system (2.21) are zero and there exists an integer $1 \leq k \leq n$ such that

$$(m+1)^i(s_n - s_n') + ... + (m+n)^i(s_1 - s_1')$$
$$= \aleph_i(s,m) - \aleph_i(s',m) = 0$$

where $i = 0, ..., k-2$ and

$$(m+1)^{k-1}(s_n - s_n') + ... + (m+n)^{k-1}(s_1 - s_1')$$

$$= \aleph_{k-1}(s,m) - \aleph_{k-1}(s',m) \neq 0$$

when $2 \leq k \leq n$, and

$$(m+1)^0(s_n - s'_n) + \ldots + (m+n)^0(s_1 - s'_1)$$

$$= \aleph_0(s,m) - \aleph_0(s',m) \neq 0$$

when $k = 1$. The integer k specifies the nonzero equation of (2.21) with a minimum power of integers $m+1, \ldots, m+n$.

This implies that

$$\aleph_0(s,m) = \aleph_0(s',m), \ldots, \aleph_{k-2}(s,m) = \aleph_{k-2}(s',m),$$

$$\aleph_{k-1}(s,m) \neq \aleph_{k-1}(s',m)$$

when $2 \leq k \leq n$ and

$$\aleph_0(s,m) \neq \aleph_0(s',m)$$

when $k = 1$. Then, using the claim with (2.19) and (2.20), we have

$$\vartheta_1(s) = \vartheta_1(s'), \ldots, \vartheta_{k-1}(s) = \vartheta_{k-1}(s'),$$

$$\vartheta_k(s) \neq \vartheta_k(s')$$

when $2 \leq k \leq n$ and

$$\vartheta_1(s) \neq \vartheta_1(s')$$

when $k = 1$.

Therefore, a sequence $s \in I_n$ can be completely specified with respect to another sequence $s' \in I_n$ by at most n of its structural numbers.

Recalling Definition 2.2, we conclude that

$$C(s, s', n) \leq n.$$

This completes the proof of the theorem.

As a result of Theorem 2.1 we see that numbers

$$\aleph_{i-1}(s,m), \quad i = 1, \ldots, n$$

can be equivalently considered instead of structural numbers

$$\vartheta_i(s), \quad i = 1, \ldots, n.$$

In particular, structural numbers and numbers (2.13) can be seen as two related sets of coordinates in the description of sequences. The merit of structural numbers comes from the geometrical interpretation.

Our main interest in Theorem 2.1 is due to the following fact established during the course of its proof. Namely, we have shown that iff

$$k = C(s, s', n) \geq 2$$

then we have (2.19) and (2.20), which by taking numbers $\aleph_i(s', m), i = 0, ..., k-1$ from the right-hand side to the left-hand side of (2.19) and (2.20) can be brought to a system of $(k-1)$ linear equations in integers

$$(s_1 - s_1')(m+n)^0 + ... + (s_n - s_n')(m+1)^0 = 0$$

$$\cdot \qquad \cdot \qquad \cdot \qquad \cdot$$

$$(s_1 - s_1')(m+n)^{k-2} + ... + (s_n - s_n')(m+1)^{k-2} = 0 \qquad (2.24)$$

and linear inequality in integers

$$(s_1 - s_1')(m+n)^{k-1} + ... + (s_n - s_n')(m+1)^{k-1} \neq 0. \qquad (2.25)$$

We will see further on that the system (2.24) and inequality (2.25) can be distinguished from a system of linear equations in integers

$$(s_1' - s_1)(m+n)^0 + ... + (s_n' - s_n)(m+1)^0 = 0$$

$$\cdot \qquad \cdot \qquad \cdot \qquad \cdot$$

$$(s_1' - s_1)(m+n)^{k-2} + ... + (s_n' - s_n)(m+1)^{k-2} = 0 \qquad (2.26)$$

and linear inequality in integers

$$(s_1 - s_1')(m+n)^{k-1} + ... + (s_n - s_n')(m+1)^{k-1} \neq 0, \qquad (2.27)$$

obtained by taking numbers $\aleph_i(s, m), i = 0, ..., k-1$ from the left-hand side to the right-hand side of (2.19) and (2.20).

The system (2.26) and inequality (2.27) are connected with the condition $k = C(s', s, n)$. Algebraically they are the same as the system (2.24) and inequality (2.25), but they are different in our terms because they represent different hierarchical formations in the structure.

Note that (2.24) and (2.25) are written somewhat differently in comparison with (2.21) and (2.22). We consider the system (2.24) as a system of integer relations, because its equations can be seen as relations between $0 \leq i \leq k-2$ powers of integers $m+1, ..., m+n$. Systems of integer relations of this kind constitute the main ingredient in the definition of the structure.

3. A NEW TYPE OF HIERARCHICAL FORMATIONS AND STRUCTURAL COMPLEXITY

There are many concepts of complexity which appear as different manifestations of intuitive notions of what the word ought to mean and there is a substantial literature of attempts to develop a theory of complexity (see for example, [3], [4], [5], [25], [26], [27], [28], [29]). However, in spite of its important role in science and powerful intuitive meanings, complexity does not have generally accepted definitions in rigorous mathematical terms.

Each of the concepts of complexity is expressed in terms of the basic concepts of a theory. Consequently, these basic concepts and the theory itself become of fundamental importance because complexity is viewed to be about general principles of self-organization in nature. However, when there is a theory describing natural phenomena, the reductionist chain of explanations always happens to appear. Namely, we have to ask why there should be such a theory and look for its explanations in terms of deeper principles. It is always the case but when a theory is final, i.e., its basic concepts are irreducible to anything more simpler. This illuminates why it is important to have a theory or a mathematical structure, whose basic concepts are irreducible.

The key idea concerning the system (2.24) and inequality (2.25) is that they represent a new type of hierarchical formations of integer relations and two-dimensional geometrical patterns. The Oxford Dictionary defines something as "complex" if it is "made of (usually several) closely connected parts" or intuitively it is understood that in order to have a complex two or more components are needed, which are joined in such a way that it is difficult to separate them.

The character of the hierarchical formations is very close to what is associated with complexity. *This motivates us to propose a concept of complexity that measures the complexity of natural systems in terms of the hierarchical formations.* The following definitions develop the concept in this direction.

Definition 2.3 (Structural Complexity C(s,s')). *Structural complexity $C(s,s')$ of a sequence $s \in I_n$ with respect to a sequence $s' \in I_n$ different from s is defined by*

$$C(s,s') = \max_{i=1,\dots,n} C(s(i), s'(i), i). \qquad (2.28)$$

In Chapter 3 we will show that structural complexity $C(s, s')$ is connected with hierarchical formations of integer relations and two-dimensional geometrical patterns. It equals the maximum level that these hierarchical formations progress to. In other words structural complexity $C(s, s')$ is the maximum level of elements that these hierarchical formations produce.

In these hierarchical formations an element of each level is composed from elements belonging the lower level. Thus the hierarchical formations, level by level, produce more complex elements until a level where no more compositions are possible. Figuratively, in the hierarchical formations elements of a level may be seen as "atoms" that combine to form "molecules" as elements belonging to the next level.

The following definition specifies the structural complexity of a sequence.

Definition 2.4 (Structural Complexity C(s)). *Structural complexity $C(s)$ of a sequence $s \in I_n$ is defined by*

$$C(s) = \max_{s' \in I_n, s \neq s'} C(s, s'). \tag{2.29}$$

Note that the definition is dependent on I_n, as structural complexity $C(s)$ is dependent on the condition $s \in I_n$. Structural complexity $C(s)$ among hierarchical formations associated with a sequence s focuses on those of them that generate elements belonging to the maximum level. Through structural complexity $C(s)$, $s \in I_n$ sequences in I_n are quantified by the maximum level of elements produced by hierarchical formations associated with the sequence s.

To appreciate the definition it may be helpful to resort to the previous analogy. In its terms, some sequences can be associated with "atoms" because related hierarchical formations produce them. Similarly, other sequences can be associated with "molecules". As "atoms" combine into "molecules", in which the "atoms" act together, the "molecules" intuitively are considered more complex then the "atoms".

This translates back to sequences. Sequences associated with "molecules" can be considered more complex than those associated with "atoms". From the structural complexity point of view, sequences can be seen as codes of the hierarchical formations, much as DNA is viewed as a code of an organism's formation.

It is also useful to put structural complexities $C(s, s')$ and $C(s)$ off the hierarchical formations' context. For example Definition 2.4 reads that $C(s)$ of a sequence $s = s_1 ... s_n \in I_n$ is the minimum number of sequence's $s(i) = s_1...s_i \in I_i$ successive structural numbers by which the sequence $s(i)$ can be completely specified in $I_i, i = 1, ..., n$.

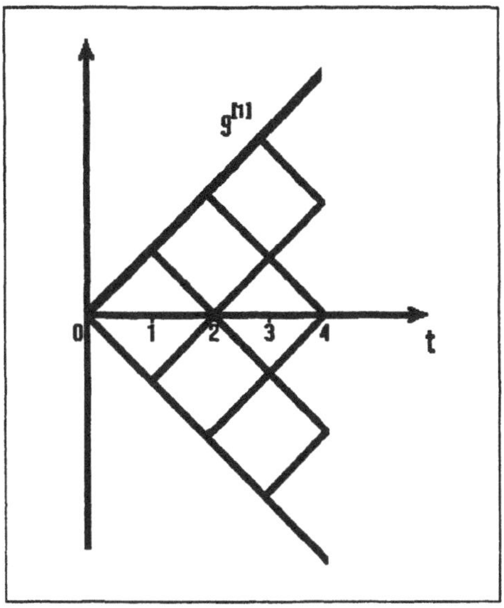

Figure 10: Considering first structural numbers of sequences in B_4 we have the well-known representation of the random walks, where the sequence s is uniquely specified by the value of a corresponding function $g^{[1]}$ at the point $t_4 = 4$.

We consider a simple example with two sequences

$$s = s_1 s_2 s_3 s_4 = +1 + 1 + 1 + 1 \in B_4,$$

$$s' = s'_1 s'_2 s'_3 s'_4 = +1 - 1 - 1 + 1 \in B_4$$

to illustrate that structural complexity measures sequences in a certain way. The example helps to realize what we mean by complete specification of a sequence in terms of structural numbers.

Let us observe that the sequence s can be completely specified in B_4 by only one structural number $\vartheta_1(s)$, whereas the sequence s' needs at least three structural numbers $\vartheta_1(s'), \vartheta_2(s'), \vartheta_3(s')$ to be in the same situation. We use the geometrical interpretation of structural numbers to prove the observation.

For the sequence s we graph functions $f^{[1]}$ as in Figure 10 for all sequences $s'' \in B_4$ such that $f = \rho_{011}(s'')$. As a result we have a well-known representation of the random walks with a time period of four steps.

By using the geometrical interpretation, we see from Figure 10 that

$$\vartheta_1(s) = g^{[1]}(t_4) > f^{[1]}(t_4) = \vartheta_1(s''),$$

where $g = \rho_{011}(s)$. This means that no sequence $s'' \in B_4$ other than the sequence s has $\vartheta_1(s)$ as its first structural number. Therefore, if $\vartheta_1(s)$ is known, then the sequence $s \in B_n$ can be completely specified by $\vartheta_1(s)$ in B_4. By considering segments $s(i), i = 1, 2, 3$ of the sequence s, it can be easily checked that

$$C(s) = 1. \tag{2.30}$$

The situation with the sequence s' is as follows. From (2.1) we get $\vartheta_1(s') = 0$ and can find that there are also five sequences in B_4 with the same first structural number. These sequences are

$$+1+1-1-1, \quad -1-1+1+1, \quad +1-1+1-1,$$

$$-1+1-1+1, \quad -1+1+1-1. \tag{2.31}$$

This fact can be made immediate in Figure 11 as well as its interpretation in terms of a global property that these sequences have in common. Namely, the global behaviour of a function in Figure 11 must be consistent with the fact that its value at the point $t_4 = 4$ equals 0. This means that the number of $+1$'s is equal to the number of -1's and the same for all the sequences (2.31). Thus $\vartheta_1(s')$ alone is not sufficient to completely specify the sequence s' in B_4, since sequences (2.31) have the same first structural number.

We can proceed then with the specification by considering the second structural number of the sequence s' in addition to the first one. At this step from (2.1) we have $\vartheta_2(s') = 0$ and the geometrical interpretation, by the integration of functions in Figure 11, helps to find out that there is only one sequence $s'' = -1+1+1-1$, which has the same second structural number. The resulting functions are pictured schematically in Figure 12. Since

$$\vartheta_1(s') = \vartheta_1(s''), \quad \vartheta_2(s') = \vartheta_2(s''), \tag{2.32}$$

it follows that the sequence s' can not be completely specified in B_4 by structural numbers $\vartheta_1(s'), \vartheta_2(s')$ only.

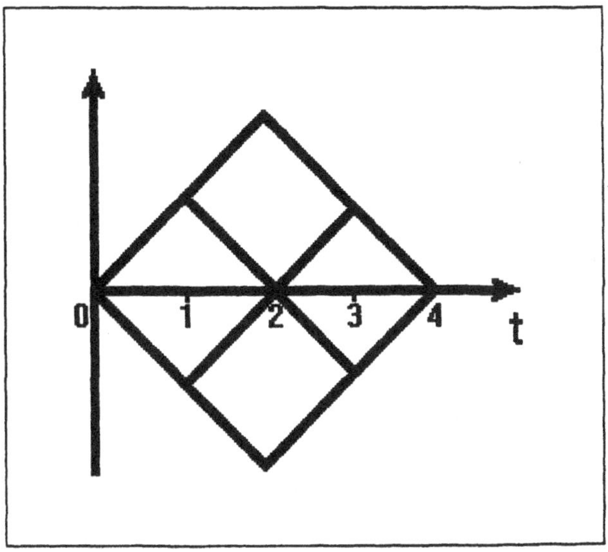

Figure 11: The consideration of sequences in B_4 whose first structural number equals 0 leads to an ensemble of the random walks that after four steps return to 0.

Condition (2.32) implies that sequences s', s'' share two global properties in common. In addition to the first property mentioned above, the second property is that the number of $(+1-1)$'s is equal to the number of $(-1+1)$'s and the same for the sequences s', s''.

The next step in the specification takes into account the third structural number $\vartheta_3(s')$ in order to distinguish the sequence

$$s' = +1 - 1 - 1 + 1$$

from the sequence

$$s'' = -1 + 1 + 1 - 1.$$

The result of the integration of functions in Figure 12 that correspond to sequences s', s'' is sketched in Figure 13 and can be used to obtain that

$$\vartheta_3(s') > \vartheta_3(s'').$$

Note that in these integrations we are allowed to think geometrically without going into numerical details and display only rough sketches of the graphs to preserve their mutual symmetrical features.

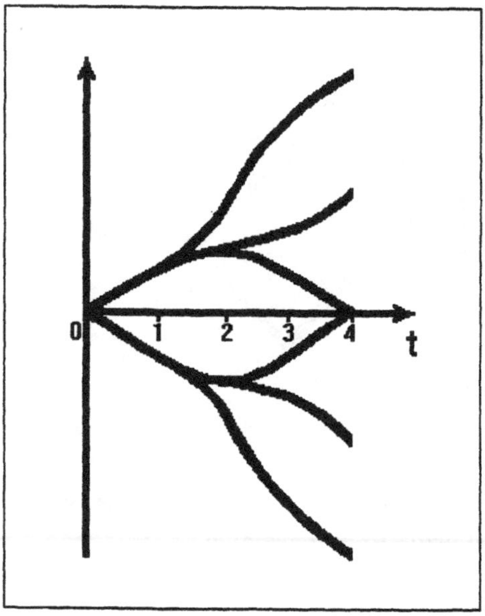

Figure 12: The figure schematically displays results of integrations of functions in Figure 11. Note that what we observe in the figure is topologically self-similar to a part in Figure 10.

Lastly $\vartheta_3(s')$ and $\vartheta_3(s'')$ are different. Thus the sequence s' can be completely specified in B_4 by three structural numbers

$$\vartheta_1(s'), \quad \vartheta_2(s'), \quad \vartheta_3(s'),$$

since no other sequence in B_4 has the same structural numbers. By considering segments $s'(i), i = 1, 2, 3$ of the sequence s', it can be easily shown that

$$C(s') = 3. \tag{2.33}$$

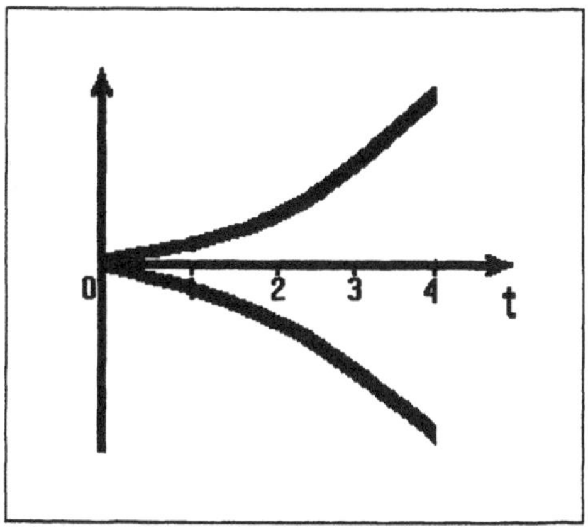

Figure 13: The figure schematically shows third structural numbers of sequences $s' = +1 - 1 - 1 + 1$, $s'' = -1 + 1 + 1 - 1$. It allows us to conclude that they are different by considering values of the functions at the point $t_4 = 4$.

Since

$$C(s) = 1 \neq 3 = C(s'),$$

we see that structural complexity can be used as a way to quantify sequences.

Remark 2.1 Notice that $C(s, s', n)$ gives two convenient ways of looking at sequences. The first way is to group sequences into families whose members are distinct in one respect, but are all alike in others.

In particular, if

$$C(s, s', n) = k \geq 2$$

then sequences s, s' belong to a family whose elements have the same values

$$\vartheta_1(s) = \vartheta_1(s'), ..., \vartheta_{k-1}(s) = \vartheta_{k-1}(s')$$

of their first $(k-1)$ structural numbers. These numbers can be used to characterize the family.

The second way is to say that a pair of sequences is closer together than another one in terms of a global measure of proximity, since $C(s, s', n)$ deals with global properties of sequences. Namely, a pair of sequences s, s' can be viewed to be closer together than a pair s, s'' if

$$C(s, s', n) > C(s, s'', n).$$

A correlation between this global measure and a local measure of proximity, usually used in Cartesian space to show how sequences are close to each other, is helpful. For

$$s = s_1...s_n, \ s' = s'_1...s'_n \in I_n$$

let

$$L(s, s', n) = \#\{j : s_j = s'_j, \ j = 1, ..., n\} \tag{2.34}$$

be a measure of proximity between sequences $s, s' \in I_n$ counting the number of components that the sequences s, s' have in common. This measure $L(s, s', n)$ is local as it counts how many times sequences s, s' locally are the same. Let

$$L(s, s') = \max_{i=1,...,n} L(s(i), s'(i), i). \tag{2.35}$$

It is easy to see that

$$L(s, s') = L(s, s', n).$$

Note that $L(s, s', n)$ can not be used, as we use $C(s, s', n)$ for (2.28) and (2.29), to measure sequences as such. Namely, a sequence $s \in I_n$ is completely specified by the components in I_n iff all n of them are known, because there are sequences in I_n that share with the sequence s all but one component, i.e.,

$$\max_{s' \in I_n} L(s, s', n) = n - 1.$$

Therefore, $L(s, s', n)$ does not allow us to measure sequences in terms of the minimum number of components needed to completely specify the sequence, contrary to what structural complexity does in the last example in (2.30) and (2.33).

Similarly, structural complexity $C(s, s', n)$ can be seen as a measure of proximity between sequences s, s' that counts the number of coordinates of a space, i.e., structural numbers, at which the sequences s, s' are equal. A correlation between these two measures $L(s, s', n)$ and $C(s, s', n)$ can be illustrated with two sequences

$$\eta(8) = +1 - 1 - 1 + 1 - 1 + 1 + 1 - 1,$$

$$\bar{\eta}(8) = -1 + 1 + 1 - 1 + 1 - 1 - 1 + 1, \qquad (2.36)$$

which are the initial segments of length 8 of the Prouhet-Thue-Morse (PTM) sequence starting with +1 and −1 respectively. The PTM sequence with respect to structural complexity will be considered in the next section.

These two sequences $\eta(8), \bar{\eta}(8)$ are quite far apart by the local measure in Cartesian space-time, since they never agree,

$$L(\eta(8), \bar{\eta}(8), 8) = \min_{s', s'' \in B_8} L(s', s'', 8) = 0,$$

where

$$\max_{s, s' \in B_8, s \neq s'} L(s, s', 8) = 7,$$

while by the global measure, as it is possible to check by computations, the sequences $\eta(8), \bar{\eta}(8)$ are close together,

$$C(\eta(8), \bar{\eta}(8), 8) = C(\eta(8), \bar{\eta}(8)) = \max_{s \in B_8} C(s) = 4,$$

where

$$\min_{s', s'' \in B_8} C(s', s'') = 1.$$

Thus we have two measures of proximity between sequences such that in the first, local one, the sequences $\eta(8), \bar{\eta}(8)$ are far apart while in the second, global one, they are close together. In other words, despite that the sequences $\eta(8), \bar{\eta}(8)$ locally have nothing in common, globally they are almost the same.

Remark 2.2 We indicate an analogy between structural complexity and some themes in algebraic geometry to relate it with another interesting context. For simplicity, let

$$s = s_1 ... s_n \in B_n, \quad s_j = \pm 1, \quad j = 1, ..., n$$

then powers

$$(m+n)^i, ..., (m+1)^i, \quad i = 0, 1, ...$$

of integers

$$m+n, ..., m+1$$

are included in (2.1) with + or − sign dependent on an associated component of the sequence. Thus, a binary sequence $s = s_1...s_n \in B_n$ gives rise to a partition of the set

$$\{m+1, ..., m+n\}$$

into two subsets. Let

$$x_j \in \{m+1, ..., m+n\}, \quad y_j \in \{m+1, ..., m+n\}, \quad 1 \leq j < n$$

denote an integer that is the jth (counting from left) integer included in (2.1) with + and − sign respectively. Then, by using this notation and gathering terms, numbers $\aleph_i(s,m), i = 0, 1, ...$ in (2.13) become

$$\aleph_i(s,m) = x_1^i + ... + x_p^i - y_1^i - ... - y_q^i = \sum_{j=1}^{p} x_j^i - \sum_{j=1}^{q} y_j^i, \qquad (2.37)$$

where $1 \leq p, q < n$ and $p + q = n$.

Firstly, the analogy comes by the fact. Let $\mathbf{X}$ be a smooth projective curve over the field $\mathbf{F_p}$, and let $\mathbf{N}_i$ be the number of $\mathbf{X}$ points over the field $\mathbf{F_q}, \mathbf{q} = \mathbf{p}^i, i = 1, 2, ...$. This information is encoded in the zeta-function

$$\zeta(\mathbf{X}, t) = \exp(\sum_{i=1}^{\infty} \mathbf{N}_i t^i / i)$$

of $\mathbf{X}$.

It is known that the zeta-function is rational if and only if there exist complex numbers

$$\mathbf{x}_j, \ j = 1, ..., p, \quad \mathbf{y}_j, j = 1, ..., q$$

such that [39]

$$\mathbf{N}_i = \sum_{j=1}^{p} \mathbf{x}_j^i - \sum_{j=1}^{q} \mathbf{y}_j^i \qquad (2.38)$$

for $i = 1, 2, ...$. Thus the rationality of the zeta-function is completely determined by the character of numbers $\mathbf{N}_i, i = 1, 2, ...$ given (2.38).

Clearly, the appearance of numbers $\mathbf{N}_i, i = 1, 2, ...$ in (2.38) is exactly analogous to the appearance of numbers $\aleph_i(s,m), i = 1, 2, ...$ in (2.37), if we make the right identification of symbols.

Secondly, the analogy can be traced further by the following fact. To be specific, from the Weil's conjectures [36], consequently proven by

Deligne [37], it follows that if a curve $\mathbf{X}$ over the field $\mathbf{F_p}$ has the genus $\mathbf{g}$ then numbers $\mathbf{N}_1, ..., \mathbf{N_g}$ determine numbers $\mathbf{N}_i$ for all $i > \mathbf{g}$ [38]. It is viewed that the proof of the Weil conjectures utilizes the most advanced techniques of modern algebraic geometry and represents one of the most remarkable mathematical achievements of this century [39].

We show that structural complexity $C(s)$ plays the same role with respect to the dependence of numbers $\aleph_i(s,m), i = 0, 1, ...$ as the genus $\mathbf{g}$ of a curve $\mathbf{X}$ over the field $\mathbf{F_p}$ does with respect to the dependence of numbers $\mathbf{N}_i, i = 1, 2, ...$. Note, by using (2.37) and (2.38), that numbers $\aleph_i(s,m), i = 1, 2, ...$ are of the same character as numbers $\mathbf{N}_i, i = 1, 2, ...$.

In particular, we claim that numbers

$$\aleph_0(s,m), ..., \aleph_{k-1}(s,m)$$

uniquely determine numbers $\aleph_i(s,m)$ for all $i > k-1$ when $k = C(s)$ and it is not the case when $1 \leq k < C(s)$ for $C(s) \geq 2$.

Let $k = C(s)$ and suppose that numbers

$$\aleph_0(s,m), ..., \aleph_{k-1}(s,m)$$

are known. Then by the claim in Theorem 2.1 structural numbers

$$\vartheta_1(s), ..., \vartheta_k(s)$$

are also known. By definition structural numbers

$$\vartheta_1(s), ..., \vartheta_k(s)$$

uniquely specify the sequence s in B_n when $k = C(s)$.

In its turn, given the sequence $s \in B_n$, numbers $\aleph_i(s,m)$ are completely determined by (2.13) for all $i \geq 0$. Consequently, numbers

$$\aleph_0(s,m), ..., \aleph_{k-1}(s,m)$$

uniquely specify numbers $\aleph_i(s,m)$ for all $i > k-1$ when $k = C(s)$.

We demonstrate that it is not the case when $k < C(s)$. Indeed, by definition of $C(s)$ there exists a sequence $s' \in B_n$ such that

$$\vartheta_1(s) = \vartheta_1(s'), ..., \vartheta_{C(s)-1}(s) = \vartheta_{C(s)-1}(s'),$$

$$\vartheta_{C(s)}(s) \neq \vartheta_{C(s)}(s'). \tag{2.39}$$

By using (2.39) from (2.17) and (2.18), we have

$$\aleph_0(s,m) = \aleph_0(s',m), ..., \aleph_{C(s)-2}(s,m) = \aleph_{C(s)-2}(s',m),$$

$$\aleph_{C(s)-1}(s,m) \neq \aleph_{C(s)-1}(s',m). \tag{2.40}$$

Thus we conclude that numbers

$$\aleph_0(s,m), ..., \aleph_{k-1}(s,m)$$

when $k < C(s)$ can not uniquely specify successive numbers for all $i > k-1$ since, given this information, the $(C(s)-1)$th number due to (2.40) can have at least two

$$\aleph_{C(s)-1}(s,m), \quad \aleph_{C(s)-1}(s',m)$$

different values.

4. STRUCTURAL COMPLEXITY AND SELF-ORGANIZATION

In this section we give more evidence in favour of structural complexity. In particular, a connection between structural complexity and self-organization is made. Results of this connection will be also used in Chapter 4.

The celebrated Prouhet-Thue-Morse (PTM) sequence [40], [41], [42] plays the main role in this connection. The PTM sequence appears in many contexts and can be described in different ways. It even has a philosophical flavour as it encodes the following process. The process starts with something, say $+1$, and then at the next stage it proceeds by negating its all previous history. At this stage it gets -1. Thus after two steps the history of the process is $+1-1$. Then the process continues further on, again by negating all of its previous history resulting in $-1+1$. After four steps the history of the process becomes $+1-1-1+1$. Then the process continues in a similar manner by negating the last negation.

One of the important discoveries to emerge from the study of complex systems, is that self-organization is closely associated with chaos. Many natural systems approach chaotic behaviour through period doubling (for example, [4], [43]).

It is known that the PTM sequence appears at the limit of the symbolic description of period doubling and can be associated with the resulting chaotic behaviour [44]. This role of the PTM sequences motivates us to consider it in terms of structural complexity. The idea is to probe whether structural complexity properly reflects self-organization.

The following two matters are of the main interest in this respect. Firstly, period doubling is seen intuitively as a process in which complexity increases. Secondly, it is viewed that a conspicuous feature of complex systems in nature is the trend towards an ever-greater complexity. A concept of complexity must elevate this kind of statements from a verbal description to a proper mathematical description.

In this section we show that structural complexity properly addresses these two important matters. Firstly, it turns out that as the period doubles, structural complexity increases by one. Secondly, there are enough reasons to believe that the PTM sequence has an extreme property since for each length, it has the maximum structural complexity among sequences. Therefore, supposedly for both matters structural complexity provides an adequate description of complexity.

There are two PTM sequences. One, denoted here by η, starts with $+1$; the other, denoted by $\bar{\eta}$, is the negation of η and starts with -1. For our purposes it is convenient to define the PTM sequence starting with $+1$ as follows.

Let $\eta_1 = +1$ and $\eta(n)$ be the initial segment of the sequence of length n, where $n = 2^{N-1}, N \geq 1$, then the initial segment of the sequence of length $2n$ is given by

$$\eta(2n) = \eta(n)\bar{\eta}(n).$$

The following theorem clarifies the situation with zeros of definite integrals related to the PTM sequence.

Theorem 2.2 (Zeros Related to the PTM Sequence). *Let*

$$\psi_+ = \rho_{011}(\eta(n)) \in W_{11}[t_0, t_n]$$

and $n = 2^N, N = 1, 2, \ldots$ *then*

1. *zeros*

$$\psi_+^{[r]}(t_k) = 0, \quad r = 1, \ldots, N \tag{2.41}$$

of the r*th integral of the function* ψ_+ *are located at the points*

$$t_k, \quad k = 1 * 2^r, 2 * 2^r, \ldots, 2^{N-r} * 2^r,$$

i.e., the r*th integral has* 2^{N-r} *equally spaced zeros in the interval*

$$(t_0, t_{m+n}],$$

2. *the sign of the* r*th integral of the function in the interval*

$$(2^r(i-1), 2^r i), \quad i = 1, \ldots, 2^{N-r}$$

is equal to the i*th symbol of the PTM sequence* η, *and*

3.

$$\psi_+^{[r]}(t_n) > 0, \quad r \geq N+1. \tag{2.42}$$

PROOF. We prove the first and second part of the theorem by induction on $N = 1, 2, \ldots$ as the function ψ_+ will be considered on enlarging, doubling intervals. We use the integer code series to evaluate integrals of the function.

For the base case $N = 1$ we have

$$\psi_+^{[1]}(t_2) = \alpha_{100}(2^0\eta_1 + 1^0\eta_2)$$

$$= \alpha_{100}(\eta_1 + \eta_2) = \alpha_{100}(+1 - 1) = 0$$

and clearly

$$\psi_+^{[1]}(t) > 0, \quad t \in (t_0, t_2)$$

to have the + sign of the first symbol of the PTM sequence. The claim follows for the base case.

Induction hypothesis. Assume that the first and the second part of the theorem are valid for N. We prove it for $N+1$.

Initially, we demonstrate that

$$\psi_+^{[1]}(t_{2^N} + t) = -\psi_+^{[1]}(t), \tag{2.43}$$

where $t \in [t_0, t_{2^N}]$. By the inductive assumption

$$\psi_+^{[1]}(t_{2^N}) = 0. \tag{2.44}$$

Consider values of the function at a point t_{2^r+k}

$$\psi_+^{[1]}(t_{2^r+k})$$

$$= \alpha_{100}((2^N+k)^0\eta_1 + \ldots + 1^0\eta_2) = \alpha_{100}\sum_{j=1}^{2^N+k}\eta_j$$

$$= \alpha_{100}\sum_{j=1}^{2^N}\eta_j + \alpha_{100}\sum_{j=2^N+1}^{k}\eta_j$$

$$= \alpha_{100}((2^N)^0\eta_1 + \ldots + 1^0\eta_{2^0}) + \alpha_{100}\sum_{j=2^N+1}^{2^N+k}\eta_j$$

$$= \psi_+^{[1]}(t_{2^N}) + \alpha_{100} \sum_{j=2^N+1}^{2^N+k} \eta_j \tag{2.45}$$

and at the point t_k

$$\psi_+^{[1]}(t_k) = \alpha_{100}(k^0\eta_1 + ... + 1^0\eta_k) = \alpha_{100} \sum_{j=1}^{k} \eta_j \tag{2.46}$$

where $k = 1, ..., 2^N$. Since (2.44), condition (2.45) implies

$$\psi_+^{[1]}(t_{2^N+k}) = \alpha_{100} \sum_{j=2^N+1}^{2^N+k} \eta_j. \tag{2.47}$$

By definition of the PTM sequence

$$\eta_{2^N+j} = -\eta_j, \quad j = 1, ..., 2^N.$$

This together with (2.46) for (2.47) gives

$$\psi_+^{[1]}(t_{2^N+k}) = \alpha_{100} \sum_{j=2^N+1}^{2^N+k} \eta_j = \alpha_{100} \sum_{j=1}^{k} -\eta_j = -\psi_+^{[1]}(t_k), \tag{2.48}$$

where $k = 1, ..., 2^N$.

Next, we show that

$$\psi_+^{[1]}(t_{2^N} + t) = -\psi_+^{[1]}(t), \tag{2.49}$$

where $t \in [t_{k-1}, t_k]$, $k = 1, ..., 2^N$. Two case are possible

1. $\eta_k = +1, \quad \eta_{2^N+k} = -1,$
2. $\eta_k = -1, \quad \eta_{2^N+k} = +1.$

In the first case we easily get

$$\psi_+^{[1]}(t) = t + \psi_+^{[1]}(t_{k-1}) - t_{k-1},$$

$$\psi_+^{[1]}(t_{2^N} + t) = -t - t_{2^N} + \psi_+^{[1]}(t_{2^N+k-1}) + t_{2^N} + t_{k-1}$$

$$= -t + \psi_+^{[1]}(t_{2^N+k-1}) + t_{k-1}$$

and, since (2.48), obtain (2.49).

Similarly, in the second case we find

$$\psi_+^{[1]}(t) = -t + \psi_+^{[1]}(t_{k-1}) + t_{k-1},$$

$$\psi_+^{[1]}(t_{2^N}+t) = t + t_{2^N} + \psi_+(t_{2^N+k-1}) - t_{2^N} - t_{k-1}$$
$$= t + \psi_+(t_{2^N+k-1}) - t_{k-1}$$

and, by using (2.48), get immediately (2.49). Since (2.49) is true for $t \in [t_k, t_{k+1}]$ and $k = 1, ..., 2^N$, this establishes (2.43).

Now our concern is to show that for $r = 1, ..., N$

$$\psi_+^{[r]}(t_k) = 0, \tag{2.50}$$

where

$$k = 2^N + 2^r, 2^N + 2 * 2^r, ..., 2^N + 2^{N-r} * 2^r.$$

In words this means that the rth integral of the function ψ_+ on the interval $[t_{2^N}, t_{2^{N+1}}]$ has a disposition of roots that is similar to the disposition of roots of the rth integral of the function on the interval $[t_0, t_{2^N}]$. To put it differently there is a one-to-one correspondence between the dispositions given by a map that shifts to the right each of those roots on the interval $[t_0, t_{2^N}]$ by 2^N.

The following chain of conditions is straightforward

$$\psi_+^{[r]}(\xi_r) = \int_{t_{2^N}}^{\xi_r} \psi_+^{[r-1]}(\xi_{r-1})d\xi_{r-1} + \omega_r, \quad t_{2^N} \leq \xi_{r-1} \leq \xi_r \leq t_{2^{N+1}},$$

$$. \quad . \quad . \quad . \quad .$$

$$\psi_+^{[2]}(\xi_2) = \int_{t_{2^N}}^{\xi_2} \psi_+^{[1]}(\xi_1)d\xi_1 + \omega_1, \quad t_{2^N} \leq \xi_1 \leq \xi_2,$$

where

$$\omega_i = \psi_+^{[i]}(t_{2^N}) = 0, \quad i = 1, ..., r$$

by the inductive assumption.

Therefore, combining the chain into one condition, we have

$$\psi_+^{[r]}(\xi_r) = \int_{t_{2^N}}^{\xi_r} ... \int_{t_{2^N}}^{\xi_2} \psi_+^{[1]}(\xi_1)d\xi_1...d\xi_r. \tag{2.51}$$

Changing variables

$$\xi_i' = \xi_i - t_{2^N}, \quad i = 1, ..., r,$$

we rewrite (2.51) as

$$\psi_+^{[r]}(\xi_r) = \int_{t_0}^{\xi_r'} ... \int_{t_0}^{\xi_2'} \psi_+^{[1]}(t_{2^N} + \xi_1')d\xi_1'...d\xi_r'. \tag{2.52}$$

Since (2.43), we have for (2.52)

$$\psi_+^{[r]}(\xi_r) = -\int_{t_0}^{\xi'_r} ... \int_{t_0}^{\xi'_2} \psi_+^{[1]}(\xi'_1) d\xi'_1 ... d\xi'_r. \tag{2.53}$$

Therefore, using proper notation, we infer from (2.53)

$$\psi_+^{[r]}(t) = -\psi_+^{[r]}(t'), \tag{2.54}$$

where

$$t = t' + t_{2^N}, \quad t \in [t_{2^N}, t_{2^{N+1}}], \quad t' \in [t_0, t_{2^N}].$$

By the inductive assumption

$$\psi_+^{[r]}(t_k) = 0, \quad k = 2^r, 2 * 2^r, ..., 2^{N-r} * 2^r$$

from (2.54), we find

$$\psi_+^{[r]}(t_k) = 0, \tag{2.55}$$

where

$$k = 2^N + 2^r, 2^N + 2 * 2^r, ..., 2^N + 2^{N-r} * 2^r, \quad r = 1, ..., N,$$

and the proof of (2.50) follows.

By the inductive assumption the sign of the rth $r = 1, ..., N$ integral $\psi_+^{[r]}$ in the interval

$$(2^r(i-1), 2^r i), \quad i = 1, ..., 2^{N-r}$$

equals the ith symbol of the PTM sequence η_i. Condition (2.50) implies that the sign of the rth integral $\psi_+^{[r]}$ in the interval

$$(2^N + 2^r(i-1), 2^N + 2^r i), \quad i = 1, ..., 2^{N-r}$$

equals $-\eta_i = \bar{\eta}_i$. By definition of the PTM sequence we get

$$\eta_{2^{N-r}+i} = \bar{\eta}_i = -\eta_i, \quad i = 1, ..., 2^{N-r}.$$

Thus we obtain that the sign of the rth integral $\psi_+^{[r]}$ in the interval $(2^r(i-1), 2^r i)$ equals the ith symbol of the PTM sequence η_i, where $i = 1, ..., 2^{N+1-r}$. This proves the second part of the theorem for $r = 1, ..., N$ on the interval $[t_0, t_{2^{N+1}}]$.

In order to prove

$$\psi_+^{[N+1]}(t_{2^{N+1}}) = 0$$

we use

$$\psi_+^{[N+1]}(t_{2N+1}) = \int_{t_0}^{t_{2N+1}} \psi_+^{[N]}(t)dt$$

$$= \int_{t_0}^{t_{2N}} \psi_+^{[N]}(t)dt + \int_{t_{2N}}^{t_{2N+1}} \psi_+^{[N]}(t)dt$$

and (2.54) for $r = N$. Thus

$$\psi_+^{[N+1]}(t_{2N+1}) = \int_{t_0}^{t_{2N}} \psi_+^{[N]}(t)dt - \int_{t_0}^{t_{2N}} \psi_+^{[N]}(t)dt = 0. \tag{2.56}$$

Combining (2.55) and (2.56) gives

$$\psi_+^{[r]}(t_k) = 0,$$

where

$$r = 1, ..., N+1, \quad k = 2^r, 2 * 2^r, ..., 2^{N+1-r} * 2^r.$$

This finishes the proof of the first part (2.41) of the theorem.

We prove the third part of the theorem similarly by induction on $N = 1, 2, \ldots$.

The base case $N = 1$. It is easy to check that

$$\psi_+^{[1]}(t) > 0, \quad t \in (t_0, t_2),$$

from which the claim for the base case follow

$$\psi_+^{[r]}(t) > 0, \quad t \in (t_0, t_2],$$

and in particular

$$\psi_+^{[r]}(t_2) > 0, \quad r > 1,$$

as required.

Induction hypothesis. Assume that the claim is valid for all $r = 1, ..., N$, i.e.,

$$\psi_+^{[r]}(t) > 0, \quad t \in (t_0, t_{2^r}], \tag{2.57}$$

then we prove it for $N + 1$. Observe that (2.57) naturally gives

$$\psi_+^{[i]}(t) > 0, \quad t \in (t_0, t_{2^r}],$$

where $i > r$.

We obtain for $t \in (t_0, t_{2N}]$

$$\psi_+^{[N+1]}(t) = \int_{t_0}^{t} \psi_+^{[N]}(\xi)d\xi > 0, \tag{2.58}$$

since by the inductive assumption

$$\psi_+^{[N]}(\xi) > 0, \quad \xi \in (t_0, t_{2^N}]. \tag{2.59}$$

For $t \in (t_{2^N}, t_{2^{N+1}}]$, by using

$$\psi_+^{[N+1]}(t_{2^{N+1}}) = 0,$$

we arrive at

$$\psi_+^{[N+1]}(t_{2^{N+1}}) = \int_{t_0}^{t} \psi_+^{[N]}(\xi)d\xi + \int_{t}^{t_{2^{N+1}}} \psi_+^{[N]}(\xi)d\xi$$

$$= \psi_+^{[N+1]}(t) + \int_{t}^{t_{2^{N+1}}} \psi_+^{[N]}(\xi)d\xi = 0.$$

In light of (2.54), we further obtain

$$\psi_+^{[N+1]}(t) = -\int_{t}^{t_{2^{N+1}}} \psi_+^{[N]}(\xi)d\xi = \int_{t'}^{t_{2^N}} \psi_+^{[N]}(\xi)d\xi, \tag{2.60}$$

where $t' = t - 2^N$ and $t' \in (t_0, t_{2^N}]$.

Since

$$\psi_+^{[N]}(\xi) > 0, \quad \xi \in (t_0, t_{2^N}],$$

then

$$\int_{t'}^{t_{2^N}} \psi_+^{[N]}(\xi)d\xi > 0,$$

and by using (2.60) with $t = t' + 2^N$, we immediately arrive at

$$\psi_+^{[N+1]}(t) > 0, \quad t \in (t_{2^N}, t_{2^{N+1}}). \tag{2.61}$$

Combining the last result with (2.59) gives

$$\psi_+^{[i]}(t) > 0, \quad t \in (t_0, t_{2^{N+1}}],$$

where $i > N+1$ and proves the third part (2.42) of the theorem.

From conditions (2.58) and (2.61) we conclude that the sign of the $(N+1)$th integral $\psi_+^{[N+1]}$ in the interval $(t_0, t_{2^{N+1}})$ equals the first symbol of the PTM sequence η, as required. This completes the proof of the second part of the theorem and finally the theorem itself.

Next, we consider the structural complexity of the PTM sequence. The point of this consideration is that, as believed, it provides us with the extreme case of the structural complexity.

Theorem 2.3 (Structural Complexity of the PTM Sequence). *Let* $n = 2^N, N = 1, 2, ...$ *then*

$$C(\eta(n), \bar{\eta}(n), n) = \log_2 n + 1 = N + 1. \tag{2.62}$$

PROOF. By definition for structural numbers of the PTM sequences $\eta(n), \bar{\eta}(n)$ of length n we have

$$\vartheta_k(\eta(n)) = \sum_{i=0}^{k-1} \alpha_{kmi}(n^i\eta_1 + (n-1)^i\eta_2 + ... + 1^i\eta_n)$$

$$= -\sum_{i=0}^{k-1} \alpha_{kmi}(n^i\bar{\eta}_1 + (n-1)^i\bar{\eta}_2 + ... + 1^i\bar{\eta}_n) = -\vartheta_k(\bar{\eta}(n)), \tag{2.63}$$

where $\delta = 1, \varepsilon = 1$ and $k = 1, 2, ...$.

From (2.63) by using the geometrical interpretation of structural numbers

$$\psi_+^{[k]}(t_n) = \vartheta_k(\eta(n)) = -\vartheta_k(\bar{\eta}(n)) = -\psi_-^{[k]}(t_n)$$

and Theorem 2.2 we have

$$\vartheta_k(\eta(n)) = \vartheta_k(\bar{\eta}(n)) = 0, \quad k = 1, ..., N \tag{2.64}$$

and

$$\vartheta_{N+1}(\eta(n)) > 0, \quad \vartheta_{N+1}(\bar{\eta}(n)) < 0.$$

Thus

$$\vartheta_{N+1}(\eta(n)) \neq \vartheta_{N+1}(\bar{\eta}(n)) \tag{2.65}$$

By definition from (2.64) and (2.65) we conclude

$$C(\eta(n), \bar{\eta}(n), n) = \log_2 n + 1 = N + 1.$$

This establishes the theorem.

Remarkable properties of the PTM sequence and numerous computational facts make it possible to formulate the following

Conjecture 2.1 (Extreme Property of the PTM Sequence) *Let*

$$s \in B_n, \quad s \neq \eta(n), \bar{\eta}(n), \quad n = 0, 1, ...$$

Then

$$C(s, \bar{s}) < C(\eta(n), \bar{\eta}(n)) = log_2 n + 1 = N + 1, \tag{2.66}$$

when $n = 2^N, N = 1, 2, \ldots$ *and*

$$C(s, \bar{s}) \leq C(\eta(n), \bar{\eta}(n)) = \lfloor log_2 n \rfloor + 1, \tag{2.67}$$

when $n \neq 2^N, N = 1, 2, \ldots$.

Moreover sequences $\eta(n), \bar{\eta}(n)$ *when* $n = 2^N, N = 1, 2, \ldots$ *attain the maximum value of structural complexity among all sequences* $s \in B_n$

$$C(\eta(n)) = C(\bar{\eta}(n)) = \max_{s \in B_n} C(s) = log_2 n + 1 = N + 1,$$

$$C(\eta(n)) = C(\bar{\eta}(n)) > C(s), \quad s \in B_n, \quad s \neq \bar{\eta}(n), \eta(n).$$

The above result holds that the PTM sequences $\eta, \bar{\eta}$ possess the maximum structural complexity. For example, the PTM sequences has the maximum structural complexity with respect to its negation than any other sequence has when $n = 2^N, N = 1, 2, \ldots$ and not less for the others n values.

For $n = 2^N, N = 1, 2, \ldots$ sequences $\tau(n)$ and $\bar{\tau}(n)$ have

$$\cdot \qquad C(\tau(n), \bar{\tau}(n)) - 1$$

equal structural numbers.

It is interesting to observe what happens with system (2.24) and inequality (2.25) when they are applied to the PTM sequences.

In particular, for

$$k = C(\eta(n), \bar{\eta}(n), n) = \log_2 n + 1 = N + 1,$$

$$n = 2^N, \quad N = 1, 2, \ldots$$

by using

$$\eta_i - \bar{\eta}_i = 2\eta_i, \quad i = 1, \ldots, n,$$

with 2 canceled out, we can rewrite (2.24) and (2.25) as

$$\eta_1 (m + n)^0 + \ldots + \eta_n (m + 1)^0 = 0$$

$$\cdot \qquad \cdot \qquad \cdot \qquad \cdot$$

$$\eta_1 (m + n)^{N-1} + \ldots + \eta_n (m + 1)^{N-1} = 0 \tag{2.68}$$

and

$$\eta_1 (m + n)^N + \ldots + \eta_n (m + 1)^N \neq 0. \tag{2.69}$$

The system (2.68) represents relations between powers of successive integers. They are included in these relations with a sign specified by a corresponding symbol of the PTM sequence. The relations can be easily converted into Prouhet identities [40], which have a formal generating procedure but lack any geometrical interpretation. Inequality (2.69) may be viewed as if relationships between the integers are exhausted.

The system (2.68) and inequality (2.69) look more impressive for concrete integers. For example, when $n = 16, m = 0$ we have a system of remarkable relations between integers $16, ..., 1$

$$16^0-15^0-14^0+13^0-12^0+11^0+10^0-9^0-8^0+7^0+6^0-5^0+4^0-3^0-2^0+1^0 = 0$$

$$16^1-15^1-14^1+13^1-12^1+11^1+10^1-9^1-8^1+7^1+6^1-5^1+4^1-3^1-2^1+1^1 = 0$$

$$16^2-15^2-14^2+13^2-12^2+11^2+10^2-9^2-8^2+7^2+6^2-5^2+4^2-3^2-2^2+1^2 = 0$$

$$16^3-15^3-14^3+13^3-12^3+11^3+10^3-9^3-8^3+7^3+6^3-5^3+4^3-3^3-2^3+1^3 = 0.$$

and an inequality between integers $16, ..., 1$

$$16^4-15^4-14^4+13^4-12^4+11^4+10^4-9^4-8^4+7^4+6^4-5^4+4^4-3^4-2^4+1^4 \neq 0$$

The last system can be written as a system of Prouhet identities

$$1^0+4^0+6^0+7^0+10^0+11^0+13^0+16^0 = 2^0+3^0+5^0+8^0+9^0+12^0+14^0+15^0$$

$$1^1+4^1+6^1+7^1+10^1+11^1+13^1+16^1 = 2^1+3^1+5^1+8^1+9^1+12^1+14^1+15^1$$

$$1^2+4^2+6^2+7^2+10^2+11^2+13^2+16^2 = 2^2+3^2+5^2+8^2+9^2+12^2+14^2+15^2$$

$$1^3+4^3+6^3+7^3+10^3+11^3+13^3+16^3 = 2^3+3^3+5^3+8^3+9^3+12^3+14^3+15^3. \tag{2.70}$$

The last inequality expresses itself as

$$1^4+4^4+6^4+7^4+10^4+11^4+13^4+16^4 \neq 2^4+3^4+5^4+8^4+9^4+12^4+14^4+15^4. \tag{2.71}$$

The system (2.70) and inequality (2.71) give us a way to demonstrate the relevance of the notion of structural complexity to complexity. We may say that the complexity of the PTM sequence of length 16 is quite enough to generate (2.70), which is intuitively perceived as a complex phenomenon. At the same time it may be said that this complexity is not enough, as we observe in (2.71), to make this phenomenon by one unit more complex.

5. INTEGER PATTERNS: VISUALISATION OF INTEGER RELATIONS

In section 3 we have shown that a complete specification of a sequence with respect to another one was reduced to a system of integer relations (2.24) and inequality (2.25). Integer and integer relation are an integral part of our mental equipment, but the scope of humans to fully appreciate integer relation is quite limited. Integer relation is but a set of abstract symbols subject to operational rules of arithmetic [45]. Usually, integer relations appear as solutions to Diophantine equations and questions about their existence and character are of the main interest. Probably the most well-known example of this type of understanding of integer relations is the last Fermat's theorem [46].

However, integer relations do not appear to us as rooted in firm reality and have the power to evoke in our minds images of concrete objects that can form into each other. *In this section we start to extend the scope of integer relation and develop a new perception of it as a geometric pattern.* In particular, we present two-dimensional geometrical patterns, called integer patterns. They help to unfold systems (2.24) into visible images of hierarchical formations as integer relations find a complete "incarnation" as integer patterns.

Definition 2.4 (Geometrical Pattern). *Let*

$$f : [a, b] \to \Re^1$$

be a piecewise constant or continuous function. Let

$$P(f, t) = \{(t, x) \in \Re^2, \min\{0, f(t)\} \leq x \leq \max\{0, f(t)\}\},$$

if the function f is continuous at the point $t \in [a, b] \subset \Re^1$, and

$$P(f, t) = \{(t, x) \in \Re^2,$$

$$\min\{0, f(t-0), f(t+0)\} \leq x \leq \max\{0, f(t-0), f(t+0)\}\},$$

if the function f is discontinuous at the point $t \in (a, b)$.

Moreover, if the function f is discontinuous at the point $t = a$ then

$$P(f, t)$$

$$= \{(t, x) \in \Re^2, \min\{0, f(a+0)\} \leq x \leq \max\{0, f(a+0)\}\},$$

and if the function f is discontinuous at the point $t = b$ then

$$P(f, t)$$

$$= \{(t, x) \in \Re^2, \min\{0, f(b-0)\} \leq x \leq \max\{0, f(b-0)\}\}.$$

Then the geometrical pattern associated with the function f is defined by

$$P(f, [a, b]) = \{P(f, t), t \in [a, b]\}$$

For illustration see Figure 14, where the geometrical pattern of a function f is brought out by shading.

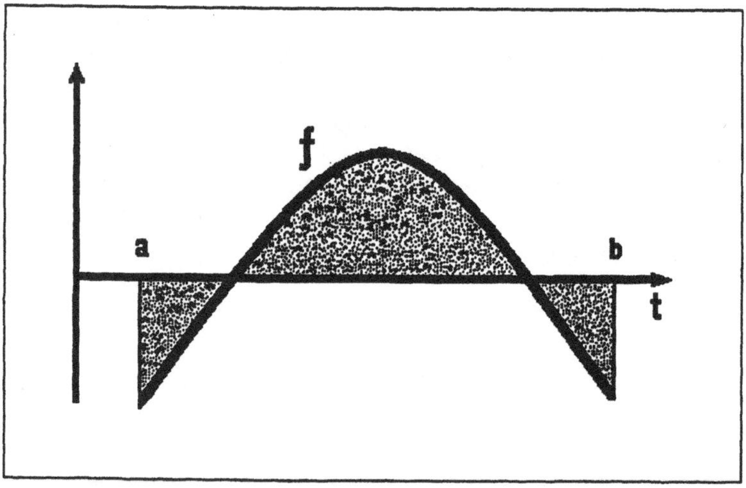

Figure 14: The pattern $P(f, [a, b])$(shaded) of a function f. The shaded area allows us to see the pattern as a spatially-localized object.

Next, we describe geometrical patterns of a special character. These patterns attract particular interest because of their connection with integers and integer relations and eventually give a new perception of integer relations as geometrical objects that can form into each other.

To be specific, set $f = f^{[0]}$ and for a function

$$f \in W_{\delta\varepsilon}[t_m, t_{m+n}]), \quad f = \rho_{m\delta\varepsilon}(s), \quad s = s_1 ... s_n \in I_n$$

let

$$T(f, [t_m, t_{m+n}])$$

be a set of intervals

$$[t_{m+i}, t_{m+i+1}] \subset [t_m, t_{m+n}]$$

such that

$$s_{i+1} \neq 0, \quad i = 0, ..., n-1.$$

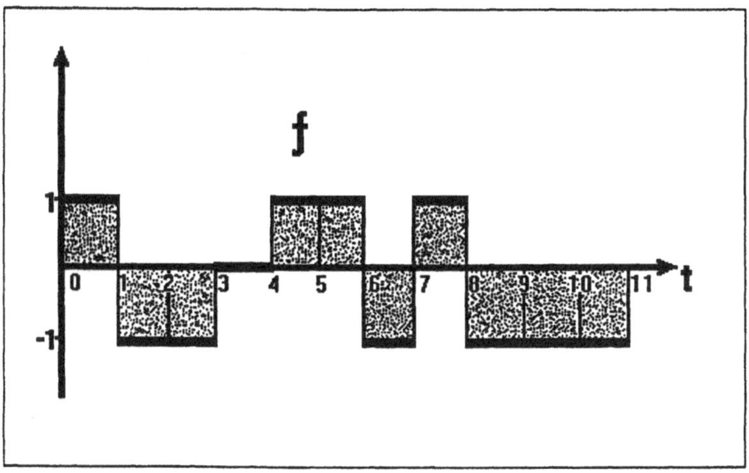

Figure 15: Integer patterns (shaded) of a function f, where $f = \rho_{011}(s)$, $s = +1-1-1\,0+1+1-1+1-1-1-1$.

Let

$$T(f^{[k]}, [t_m, t_{m+n}]), \quad k = 1, 2, ...$$

be a set of intervals

$$[t_i, t_j] \subseteq [t_m, t_{m+n}], \quad m \leq i, \quad i+1 < j, \quad j \leq m+n$$

such that

$$f^{[l]}(t_i) = 0, \quad f^{[l]}(t_j) = 0, \quad l = 1, ..., k,$$

and there exists no

$$t_{j'}, \quad j' = i+1, ..., j-1$$

such that $f^{[l]}(t_{j'}) = 0$ for all $l = 1, ..., k$ at one time.

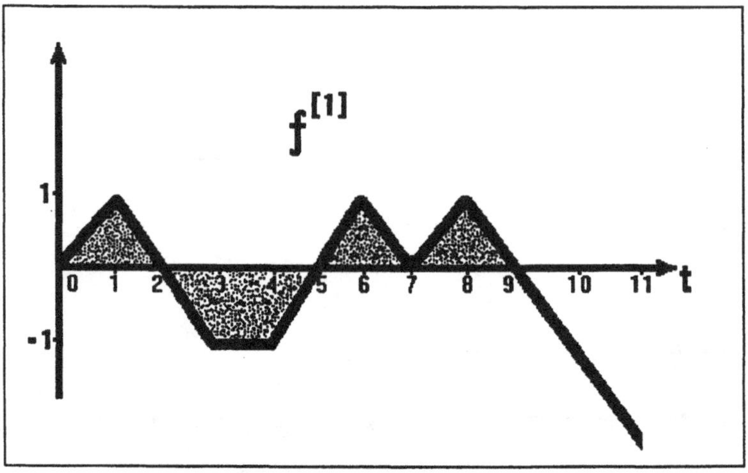

Figure 16: Integer patterns (shaded) of the first integral $f^{[1]}$ of the function f depicted in Figure 15.

Definition 2.5 (Integer Pattern). *A geometrical pattern*

$$P(f^{[k]}, [t_i, t_j]), \quad k = 0, 1, ...$$

is called an integer pattern of the kth integral of a function

$$f \in W_{\delta\varepsilon}[t_m, t_{m+n}])$$

if

$$[t_i, t_j] \in T(f^{[k]}, [t_m, t_{m+n}]) \neq \emptyset.$$

Points t_i, t_j of the integer pattern $P(f^{[k]}, [t_i, t_j])$ are called its endpoints.

For instance, Figures 15 and 16 shows integer patterns

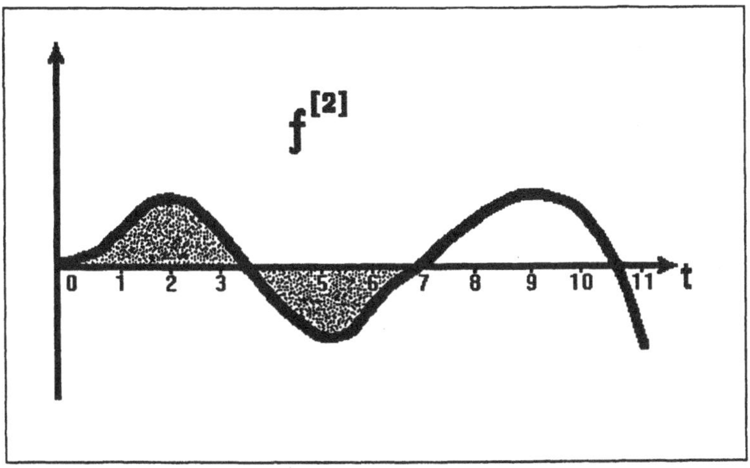

Figure 17: Shaded integer pattern $P(f^{[2]}, [t_0, t_7])$ and sketched integral $f^{[2]}$ of the function f depicted in Figure 15.

$$P(f, [t_0, t_1]), \quad P(f, [t_1, t_2]), \quad P(f, [t_2, t_3]), \quad P(f, [t_4, t_5]),$$
$$P(f, [t_5, t_6]), \quad P(f, [t_6, t_7]), \quad P(f, [t_7, t_8]), \quad P(f, [t_8, t_9]),$$
$$P(f, [t_9, t_{10}]), \quad P(f, [t_{10}, t_{11}])$$

of a function $f = \rho_{011}(s)$, where

$$s = +1 - 1 - 1\, 0 + 1 + 1 - 1 + 1 - 1 - 1 - 1 \in I_{11},$$
$$I = \{-1, 0, +1\},$$

and integer patterns

$$P(f^{[1]}, [t_0, t_2]), \quad P(f^{[1]}, [t_2, t_5]),$$
$$P(f^{[1]}, [t_5, t_7]), \quad P(f^{[1]}, [t_7, t_9])$$

of its first integral $f^{[1]}$.

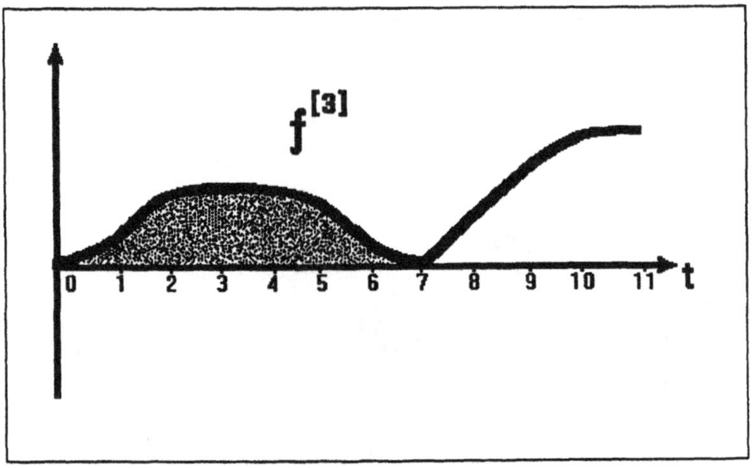

Figure 18: Shaded integer pattern $P(f^{[3]}, [t_0, t_7])$ and sketched integral $f^{[3]}$ of the function f depicted in Figure 15.

In Figures 17 and 18 an integer pattern $P(f^{[2]}, [t_0, t_7])$ of the second integral $f^{[2]}$ and an integer pattern $P(f^{[3]}, [t_0, t_7])$ of the third integral $f^{[3]}$ are brought out by shading. In the case of $f^{[2]}$ the integer pattern $P(f^{[2]}, [t_0, t_7])$ can be viewed to consist of two patterns

$$P(f^{[2]}, [0, 3.5]), \quad P(f^{[2]}, [3.5, 7]), \tag{2.72}$$

which are not integer patterns because two conditions are not satisfied, i.e., $f^{[1]}(3.5) \neq 0$ and 3.5 is not an integer. Patterns (2.72) are identical in shape and can be viewed as opposite in sign to each other.

Under the integration the integer pattern

$$P(f^{[2]}, [t_0, t_7])$$

transforms into the integer pattern

$$P(f^{[3]}, [t_0, t_7]).$$

It is tempting to picture this transformation as if patterns (2.72) merge together and form the integer pattern $P(f^{[3]}, [t_0, t_7])$.

In the next chapter we demonstrate how integer patterns can be related to integer relations and here do no more than present integer relations corresponding to integer patterns

$$P(f^{[2]}, [t_0, t_7]), \quad P(f^{[3]}, [t_0, t_7]).$$

Namely, $P(f^{[2]}, [t_0, t_7])$ corresponds to a relation between integers $1, 2, 3, 5, 6, 7$

$$+7^1 - 6^1 - 5^1 + 3^1 + 2^1 - 1^1 = 0,$$

and $P(f^{[3]}, [t_0, t_7])$ corresponds to another relation between integers $1, 2, 3, 5, 6, 7$

$$+7^2 - 6^2 - 5^2 + 3^2 + 2^2 - 1^2 = 0.$$

6. SYSTEMS OF INTEGER RELATIONS AND HIERARCHICAL FORMATIONS: PARTICULAR EXAMPLE

In this section for a particular example we show how system (2.24) and inequality (2.25) can be associated with hierarchical formations of integer relations and integer patterns. Figuratively, such examples are a window that allows us to view properties of the structure and see answers that may not be possible to obtain in the formal way. The example demonstrates nonlocal order and large symmetry in hierarchical formations. Hopefully, this gives us an insight about the harmony of the structure.

Consider a sequence $\varsigma \in B_n$ connected with the Fibonacci numbers, tiling and quasicrystals (for example, [47]). Sequences, similar to the sequence ς, are natural in the sense that they arise in many physical situations independently of the exact character of interactions (for example, [48]).

The sequence ς is formed by the substitution rules

$$1 \to 1 - 1, \quad -1 \to 1.$$

Starting with $+1$, successive substitutions produce

$$+1 - 1, \quad +1 - 1 + 1, \quad +1 - 1 + 1 + 1 - 1, \dots .$$

Let

$$s = +1 - 1 + 1 + 1 - 1 + 1 - 1 + 1 + 1 - 1 + 1 + 1 - 1 + 1 - 1$$

be the initial segment of length 15 of ς. There is a sequence

$$s' = -1 + 1 + 1 + 1 + 1 - 1 - 1 + 1 + 1 + 1 - 1 + 1 - 1 - 1 + 1$$

such that $C(s, s', 15) = 4$, i.e.,

$$\vartheta_k(s) = \vartheta_k(s'), \quad k = 1, 2, 3,$$
$$\vartheta_4(s) \neq \vartheta_4(s'), \tag{2.73}$$

where $\varepsilon = 1, \delta = 1$.

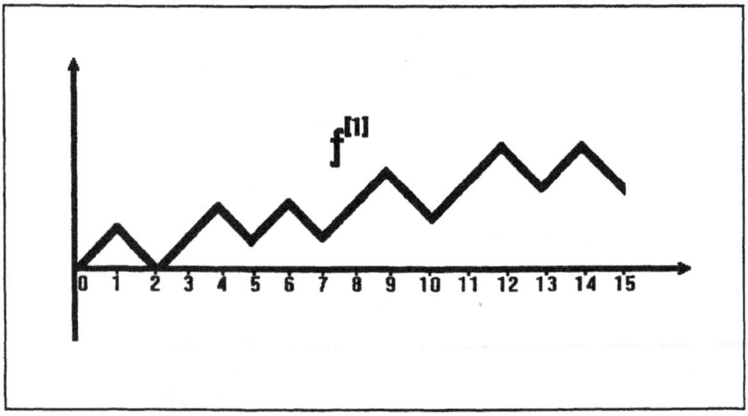

Figure 19: Graph of the function $f^{[1]}$ exhibits no visible properties when regarded by itself.

Clearly, the condition (2.73) due to Theorem 2.1 for $m = 0$ is equivalent to a system of integer relations

$$(s_1 - s_1')15^0 + \ldots + (s_{15} - s_{15}')1^0 = 0$$

$$(s_1 - s_1')15^1 + ... + (s_{15} - s_{15}')1^1 = 0$$
$$(s_1 - s_1')15^2 + ... + (s_{15} - s_{15}')1^2 = 0, \qquad (2.74)$$

and inequality

$$(s_1 - s_1')15^3 + ... + (s_{15} - s_{15}')1^3 \neq 0. \qquad (2.75)$$

We are interested to show what phenomena are connected with (2.74) and (2.75).

By using the geometrical interpretation of structural numbers we graph the first integrals $f^{[1]}, f_{\prime}^{[1]}$ of functions

$$f = \rho_{011}(s), \quad f_{\prime} = \rho_{011}(s')$$

as shown in Figures 19 and 20.

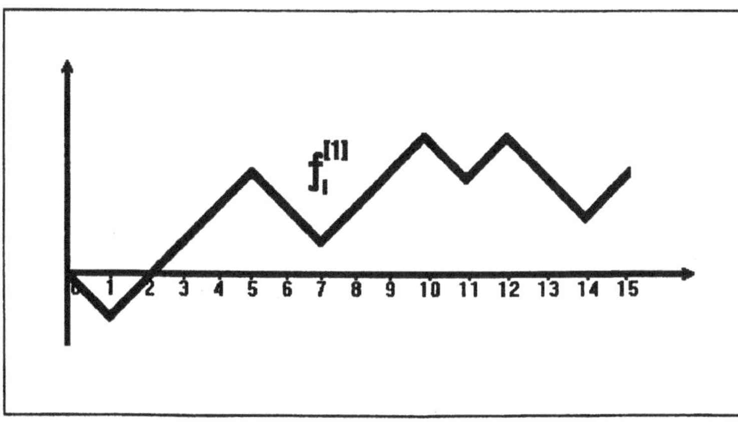

Figure 20: Graph of the function $f_{\prime}^{[1]}$ exhibits no visible properties when regarded by itself.

When regarded separately graphs in Figures 19 and 20 exhibit no visible properties, but depicted together, properties emerge as in Figure

21. These properties are manifested explicitly in the graph of a function $g^{[1]}$, where

$$g = \rho_{011}(s - s') = f - f_{\prime}$$

and $s - s'$ is a sequence $s'' = s''_1 ... s''_n$ such that (see Figure 22).

$$s''_i = s_i - s'_i, \quad i = 1, ..., n.$$

Graphs of functions $g^{[i]}, i = 0, ..., 3$ have visual symmetry and their integer patterns are brought out in Figures 22, 23, 24 and 25 by shading.

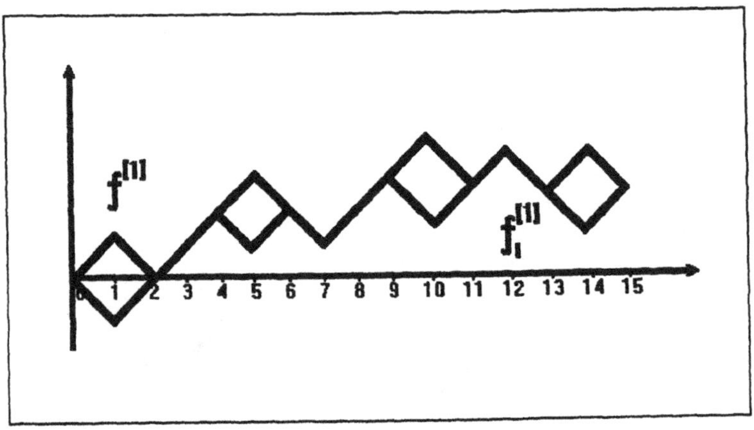

Figure 21: Graphs of functions $f^{[1]}$ and $f_{\prime}^{[1]}$ depicted together uncover visible properties.

From these figures without computations and using a symmetrical and geometrical viewpoint, it follows easily that

$$g^{[k]}(t_{15}) = 0, \ k = 1, 2, 3,$$

$$g^{[4]}(t_{15}) \neq 0.$$

Consequently

$$f^{[k]}(t_{15}) = f_{\prime}^{[k]}(t_{15}), \ k = 1, 2, 3,$$

$$f^{[4]}(t_{15}) \neq f_{i}^{[4]}(t_{15})$$

or by (2.2).

$$\vartheta_k(s) = \vartheta_k(s'), \quad k = 1, 2, 3,$$
$$\vartheta_4(s) \neq \vartheta_4(s'),$$

and this leads us exactly to (2.74) and (2.75).

Thus the system (2.74) and inequality (2.75) can be associated with a hierarchical formation of the integer patterns that is amenable to an explanation in terms of visible things.

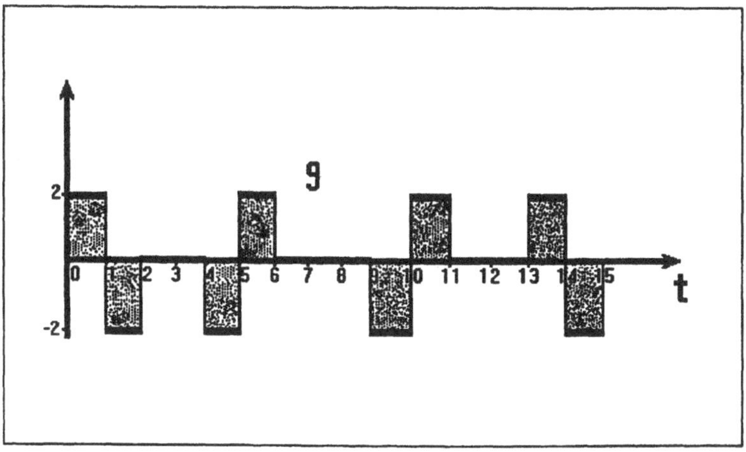

Figure 22: Graph of the function g. As we integrate, it appears as if the integer patterns merge together in pairs to form new integer patterns belonging to the next level.

In particular, Figures 22, 23, 24 and 25 may be viewed as though they were experimental results successively representing levels of this hierarchical formation. It is interesting to observe that as we integrate, it appears that corresponding integer patterns merge together in pairs to form new integer patterns belonging to the next level.

Since

$$g^{[k]}(t) > 0, \quad t \in (t_0, t_{15}], \quad k = 4, 5, \ldots$$

then there are no integer patterns belonging to the $(k+1)$th level.

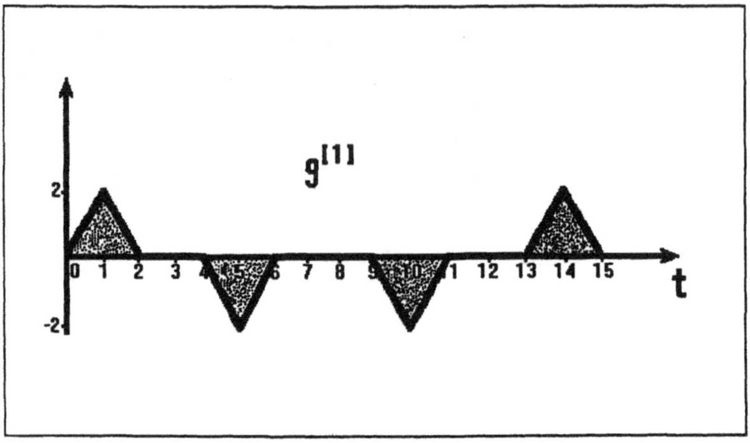

Figure 23: Graph of the first integral $g^{[1]}$ of the function g. As we integrate, it appears as if the integer patterns merge together in pairs to form new integer patterns belonging to the next level.

Moreover, the levels combined together form a unified geometric object, in which a tree diagram can be noticed (see Figure 26). This object displays many interesting features that address our insight. Firstly, in Figure 26 we immediately see a large symmetry involved in the hierarchical formation of the integer patterns. It is worthy of note that the symmetry is exhibited without any need for further explanation. The eye at once sees the crucial image that yields it.

Secondly, we witness an order involved in the hierarchical formation of these patterns. This order is very much connected with the symmetry and is of a nonlocal character. It concerns all integer patterns of a level and across all the levels simultaneously.

In particular, the order specifies how integer patterns of a level have to make the exactly right connections between themselves, in terms of their position and shape, to grow from one level to the next one. One may say that the order combine integer patterns of a level into integer patterns that they produce at the next level. The order is left very rigid because of a minor change in it leads to the collapse of the whole object.

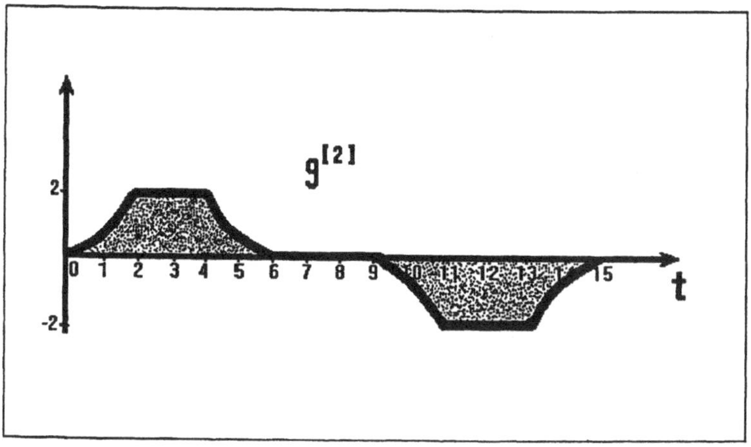

Figure 24: Graph of $g^{[2]}$ (sketched). It is tempting to view the integer patterns (shaded) as abstract entities corresponding to natural systems.

Thirdly, Figure 26 demonstrates self-similarity that accompanies the hierarchical formation. It arises as at each level, topologically, we have the same situation when integer patterns join or branch. The difference is in the scale and shape. This is inspirational, since it is well-known that a conspicuous feature of many complex systems in nature is the possession of self-similar structures [49].

Most importantly Figure 26 gives us a clear understanding that the integer patterns uncover themselves through their connections to a unified whole as its integrated parts. This whole consists of infinite hierarchical levels with interconnected integer patterns. For example, the integer pattern of the forth level, which may be viewed as the result of the hierarchical formation, is just an integrated building block of the level. This naturally motivates us to make this whole a subject of special interest.

The mathematical structure is designed in the next chapter to formally capture this whole.

We can think of the integer patterns, of each coming level, as next in order of complexity to the integer patterns of the previous level, since every integer pattern of a successive level is formed from integer patterns of the previous one.

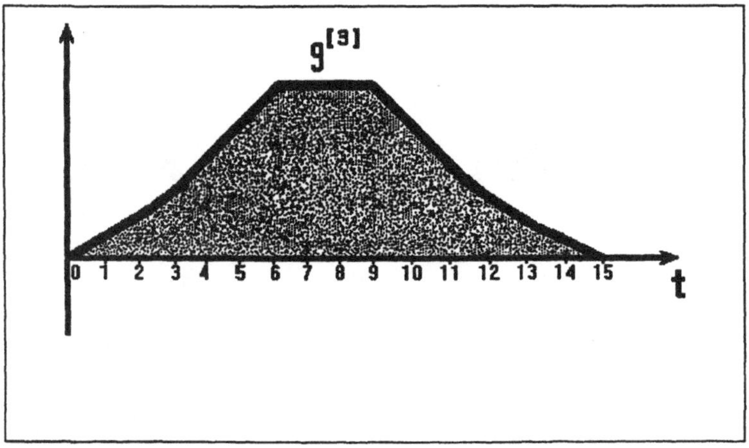

Figure 25: Graph of $g^{[3]}$ (sketched). It is tempting to view the integer pattern (shaded) as an abstract entity corresponding to a natural system.

From Figure 26 we see that condition $C(s, s', 15) = 4$ is the maximum level of the integer patterns produced by the hierarchical formation. All these four levels of the hierarchical formation contain only integer patterns, while the next levels do not have any integer patterns produced by it.

The hierarchical formation may be seen to be close to the intuitive understanding of what complexity is all about. This explains why we use the ambitious term "structural complexity" as the name for $C(s, s')$ and $C(s)$.

Next, it is possible to translate these geometrical considerations into an algebraic form. This is a place where a hierarchical formation of integer relations comes into view.

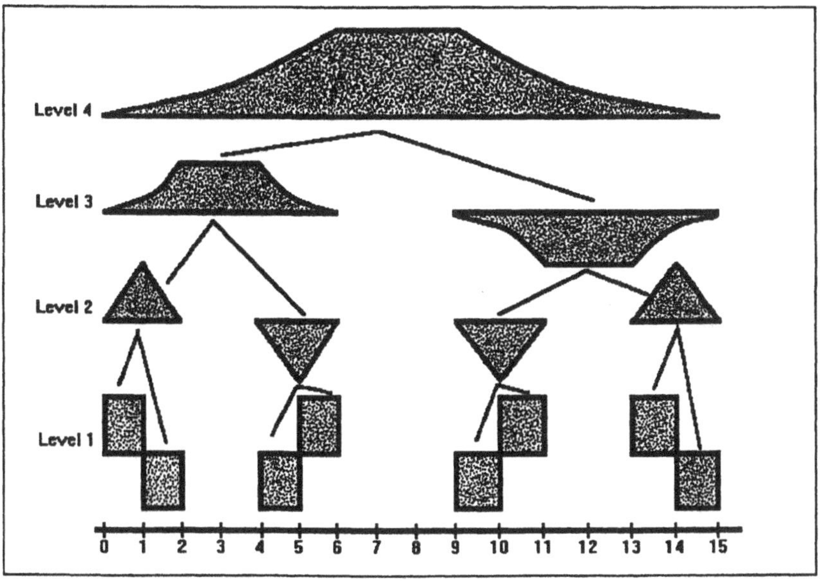

Figure 26: A hierarchical formation of integer patterns. This hierarchical formation has order and large symmetry. The order is nonlocal as it concerns all the integer patterns of a level and across all levels simultaneously. The unified geometric object of these integer patterns is invisibly controled by relations between integers $1, 2, 5, 6, 10, 11, 14, 15$. The structural complexity $C(s, s', 15) = C(s, s') = 4$ is the highest level of this hierarchical formation.

Namely, substituting $s_i, s'_i, i = 1, ..., 15$ explicitly in (2.74) and (2.75), we obtain a system of integer relations

$$+2*15^0 - 2*14^0 - 2*11^0 + 2*10^0 - 2*6^0 + 2*5^0 + 2*2^0 - 2*1^0 = 0$$

$$+2*15^1-2*14^1-2*11^1+2*10^1-2*6^1+2*5^1+2*2^1-2*1^1=0$$
$$+2*15^2-2*14^2-2*11^2+2*10^2-2*6^2+2*5^2+2*2^2-2*1^2=0$$

and inequality

$$+2*15^3-2*14^3-2*11^3+2*10^3-2*6^3+2*5^3+2*2^3-2*1^3\neq 0.$$

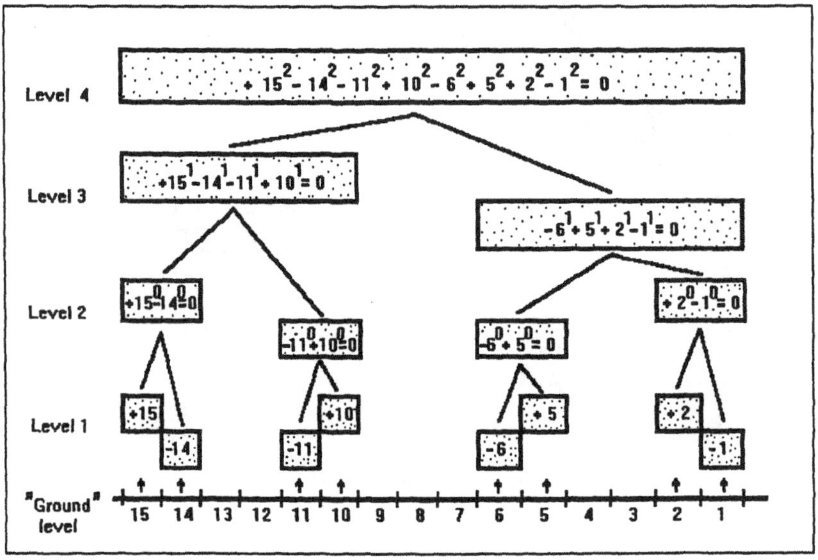

Figure 27: A hierarchical formation of integer relations. There is a one-to-one correspondence between this hierarchical formation and the hierachical formation of integer patterns in Figure 26. The figure gives us a clear understanding that everything develops from the integers, i.e., "ground" level, as one undivided whole and the hierarchical formation is its integrated part.

To have more elegance, the system and inequality can be rearranged with 2 canceled out. This process then gives a system of Prouhet's type identities

$$1^0+6^0+11^0+14^0=2^0+5^0+10^0+15^0$$

$$1^1 + 6^1 + 11^1 + 14^1 = 2^1 + 5^1 + 10^1 + 15^1$$
$$1^2 + 6^2 + 11^2 + 14^2 = 2^2 + 5^2 + 10^2 + 15^2 \quad (2.76)$$

and inequality

$$1^3 + 6^3 + 11^3 + 14^3 \neq 2^3 + 5^3 + 10^3 + 15^3 \quad (2.77)$$

between integers $1, 2, 5, 6, 10, 11, 14, 15$. Inequality (2.77) may be viewed to suggest that relationships between the integers presented in (2.76) are exhausted.

The system (2.74) viewed by itself does not show that in fact there is a hierarchical formation of these relations behind it. The geometrical interpretation of structural numbers helps us to identify this hierarchical formation shown in Figure 27.

In Figures 26 and 27 we observe a perfect match between the integer patterns and the integer relations, and that the hierarchical formation of the integer patterns corresponds to the hierarchical formation of the integer relations.

Overall, in these figures we witness a connection between two different modes of the mathematical understanding. The two modes being geometry and integers come to light in an interesting manner. In particular, this tells us that geometrical patterns may be just relations between integers and the integers themselves can be considered as the ultimate building blocks from which the hierarchical formations start.

Consequently, (2.74) and (2.75) are a mathematical form that represents two hierarchical formations, one is of geometrical integer patterns and the other is of algebraic integer relations. With this example in mind we proceed in the next chapter with formal definitions of the structure.

Chapter 3

A NEW TYPE OF HIERARCHICAL FORMATIONS AND THE STRUCTURE

In Chapter 2 for a particular example the system (2.24) *and inequality* (2.25) *was associated with a hierarchical formation, which had two distinctive forms. In the first form, the elements are integer relations suggesting that the ultimate building blocks are integers from which "everything" develops as one whole. In the second form, the elements are two-dimensional geometric patterns, i.e., the integer patterns. The example helps us to understand that the hierarchical formation constitutes one possible hierarchical formation among many others admitted by a larger structure. This naturally motivates us to consider and define this structure.*

This consideration consists of two parts. Firstly, in section 1 we associate the system (2.24) *and inequality* (2.25) *with a hierarchical formation of integer relations. Secondly, in section 2 we show that there is an isomorphism between the hierarchical formation of integer relations and a hierarchical formation of integer patterns. This helps us in section 3 to define a structure, called a web of relations, that integrates all such hierarchical formations into one whole. In section 4 we argue in favour of the web of relations to be considered as a final theory. The web of relations does not rely on information about space-time and is based on integers as the single assumption.*

1. HIERARCHICAL FORMATIONS OF INTEGER RELATIONS

In this section we associate the system integer equations (2.24)

$$(s_1 - s_1')(m+n)^0 + ... + (s_n - s_n')(m+1)^0 = 0$$

. . . .

$$(s_1 - s_1')(m+n)^{k-2} + ... + (s_n - s_n')(m+1)^{k-2} = 0 \qquad (3.1)$$

and inequality (2.25)

$$(s_1 - s_1')(m+n)^{k-1} + ... + (s_n - s_n')(m+1)^{k-1} \neq 0. \qquad (3.2)$$

with a hierarchical formation of integers and integer relations

$$\left\{ \begin{array}{c} \textit{System of Integer Equations(3.1)} \\ \textit{and Inequality(3.2)} \end{array} \right\}$$

$$\Longrightarrow \left\{ \begin{array}{c} \textit{Hierarchical Formation} \\ \textit{of Integer Relations} \end{array} \right\}$$

This hierarchical formation can be represented as a hierarchical set

$$WR(s, s', n, m, I_n)$$

of $C(s, s', n)$ levels whose elements are integers and integer relations. Elements of the set belonging to neighbouring levels are connected and we need a notion to describe the connection.

The following notation will be widely used. For sequences

$$s = s_1 ... s_n, \quad s_1' ... s_n' \in I_n$$

let $\tau(j)$ be equal i when the condition $s_i - s_i' \neq 0$ is the jth occurrence when components of the sequences are not equal as the index i takes successive values $1, ..., n$. For simplicity sequences s, s' are omitted in the notation $\tau(j)$ as the dependence will always come from a context.

Let

$$H(s, s') = \#\{j : s_j - s_j' \neq 0, j = 1, ..., n\}.$$

Set $\tau(0) = 0$ and note

$$\tau(j-1) < \tau(j), \quad j = 1, ..., H(s, s').$$

The following definition specifies the connection between elements of the structure when they are viewed as integer relations. The connection comes if elements of a level form an element of the next level.

Definition 3.1 (Composition of Integer Relations). *Let*

$$s = s_1 ... s_n, \ s_1' ... s_n' \in I_n, \quad s \neq s'$$

and

$$\sum_{j=r_{k+2,i-1}+1}^{r_{k+2,i}} (s_{\tau(j)} - s'_{\tau(j)})(m+n-\tau(j)+1)^k = 0 \tag{3.3}$$

for $i = 1, ..., N$ *be a collection of* $N \geq 1$ *integer relations, where* $k \geq 0, m$ *are integers and*

$$0 \leq r_{k+2,0}, \quad r_{k+2,i-1} < r_{k+2,i}, \quad i = 1, ..., N, \quad r_{k+2,N} \leq n.$$

If

$$\sum_{i=1}^{N} \sum_{j=r_{k+2,i-1}+1}^{r_{k+2,i}} (s_{\tau(j)} - s'_{\tau(j)})(m+n-\tau(j)+1)^{k+1}$$

$$= \sum_{j=r_{k+2,0}+1}^{r_{k+2,N}} (s_{\tau(j)} - s'_{\tau(j)})(m+n-\tau(j)+1)^{k+1} = 0, \tag{3.4}$$

with

$$\sum_{i=1}^{N'} \sum_{j=r_{k+2,i-1}+1}^{r_{k+2,i}} (s_{\tau(j)} - s'_{\tau(j)})(m+n-\tau(j)+1)^{k+1} \neq 0,$$

where $N > 1$, $N' = 1, ..., N-1$ *then the integer relation* (3.4) *is called the composition of integer relations* (3.3) *and the integer relations* (3.3) *in turn are called the decomposition of the integer relation* (3.4).

The ith $i = 1, ..., N$ integer relation of the collection (3.4) is viewed as a relation between the kth $k = 0, 1, ...$ power of integers

$$m+n-\tau(r_{k+2,i-1}+1)+1, m+n-\tau(r_{k+2,i-1}+2)+1, ...,$$

$$m+n-\tau(r_{k+2,i})+1 \tag{3.5}$$

and an element of the $(k+2)$th level of a structure.

Note that, given sequences s, s', the ith element of (3.3) is specified by integers $r_{k+2,i-1}, r_{k+2,i}$, which with the help of the function τ determines what integers (3.5) are included in its representation. In turn we see that integer relation (3.4) can be viewed as an element of the next $(k+3)$th level of the same structure specified by integers $r_{k+2,0}, r_{k+2,N}$.

Let us consider Definition 3.1 more closely. The composition (3.4) is clearly more than the simple sum of the elements (3.3)

$$\sum_{i=1}^{N} \sum_{j=r_{k+2,i-1}+1}^{r_{k+2,i}} (s_{\tau(j)} - s'_{\tau(j)})(m+n-\tau(j)+1)^k$$

$$= \sum_{j=r_{k+2,0}+1}^{r_{k+2,N}} (s_{\tau(j)} - s'_{\tau(j)})(m + n - \tau(j) + 1)^k = 0,$$

since powers of integers in (3.4) are increased by 1.

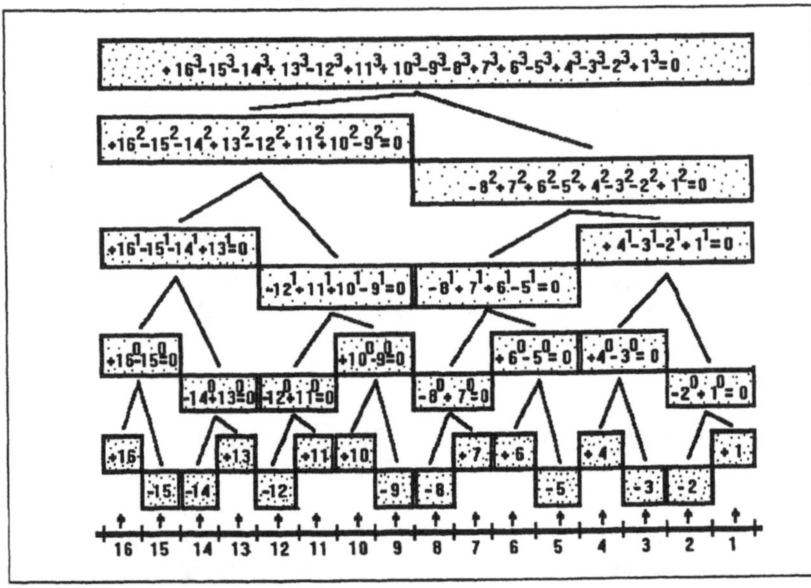

Figure 28: A hierarchical set of integer relations. Note that relations $+16^1 - 15^1 - 14^1 + 13^1 = 0$ and $-16^1 + 15^1 + 14^1 - 13^1 = 0$ are different elements

The composition (3.4) may be said to have a greater complexity than the collection (3.3), because given the composition the collection is also given. In addition, the appearance of (3.4) traces features of the elements (3.3) and it may be viewed that they are harmoniously organized into a new whole in (3.4).

Definition 3.1 captures a process in which elements become composed and this leads to the emergence of a new element. The case when $N = 1$ may be thought of as internal transformations in (3.3) and (3.4) resulting in (3.4) and (3.3) respectively.

In general, a collection of integer relations of type (3.3) does not give a composition. This means that elements (3.3) must have a special property together for the composition (3.4) to emerge. This raises two main questions.

Firstly, what is this special property that is responsible for the emergence of an element (3.4) from elements (3.3). Secondly, what is the principle that, when the property is given, organizes elements of a collection (3.3) and as the result gives rise to an element (3.4). In the next section we consider these questions by developing a new perception for the integer relations as two-dimensional geometrical patterns.

It is convenient to represent the composition of elements (3.3) by a diagram. The diagram has a "ground" level that is represented by integers associated with unit intervals in the reverse order (for example, see Figure 27 as an illustration). In the diagram the $(k+2)$th $k = 0, 1, ...$ level is linked with a horizontal line which is positioned at some distance above a horizontal line of the previous $(k+1)$th level to represent compositions across the levels. The ith $i = 1, ..., N$ element of the set is placed in an rectangle which is located at the $(k+2)$th level and whose left and right side are specified respectively by the left and right end of intervals corresponding to integers

$$m + n - \tau(r_{k+2,i-1} + 1) + 1, \quad m + n - \tau(r_{k+2,i}) + 1.$$

Depending on the sign of

$$s_{\tau(r_{k+2,i-1}+1)} - s'_{\tau(r_{k+2,i-1}+1)},$$

i.e., + or −, the rectangle in addition lies above or below the line (see Figures 27 and 28).

In the diagram element (3.4) is placed at the $(k+3)$th level in a rectangle, whose position in the similar way is specified by integers

$$m + n - \tau(r_{k+2,0} + 1) + 1, \quad m + n - \tau(r_{k+2,N}) + 1$$

and the sign of

$$s_{\tau(r_{k+2,0}+1)} - s'_{\tau(r_{k+2,0}+1)}.$$

In the diagram edges between the rectangles on the $(k+2)$th level and the rectangle on the $(k+3)$th level are used to indicate the connection between them in the composition (3.4). As a result the composition in the diagram takes the form of a tree on the $(k+2)$th and $(k+3)$th levels.

In diagrammatic language the tree reads that the composition of elements (3.3) results in the formation of the element (3.4) or the decomposition of the element (3.4) results in the formation of the elements (3.3).

Note that by our convention elements

$$\sum_{j=r_{k+2,i-1}+1}^{r_{k+2,i}} (s_{\tau(j)} - s'_{\tau(j)})(m+n-\tau(j)+1)^{k+1} = 0,$$

$$\sum_{j=r_{k+2,i-1}}^{r_{k+2,i}} -(s_{\tau(j)} - s'_{\tau(j)})(m+n-\tau(j)+1)^{k+1} = 0,$$

are depicted differently and are seen as opposite to each other, where $i = 1, ..., N$.

With the notion of composition at hand we proceed with the definition of the set

$$WR(s, s', n, m, I_n).$$

It may be helpful to refer to Figure 27 as well as Figure 28 during the definition. In Figure 28 a hierarchical set of integer relations

$$WR(\eta(16), \bar{\eta}(16), 16, 0, B_{16})$$

is shown, where

$$\eta(16)$$
$$= +1-1-1+1-1+1+1-1-1+1+1-1+1-1-1+1$$

is the initial segment of length 16 of the PTM sequence η and for convenience common multiple 2 is omitted in the elements' representation.

The set $WR(s, s', n, m, I_n)$ takes an interesting interpretation when a part of the "ground" level, i.e., integers

$$m+n, m+n-1, ..., m+1, \qquad (3.6)$$

is used in its description. These elements of the level are numbered in the reverse order, i.e., the first element is $m+n$ and the last element is $m+1$. It is interesting to consider that ground level can "generate" integers and this leads to the formation of elements of the first level.

The ith element $i = 1, ..., N_1$ of the set belonging to the first level is

$$(s_{\tau(i)} - s'_{\tau(i)})(m+n-\tau(i)+1), \qquad (3.7)$$

where $N_1 = H(s, s')$ and is viewed to be composed of

$$|s_{\tau(i)} - s'_{\tau(i)}| \neq 0$$

integers

$$m + n - \tau(i) + 1$$

generated from the "ground" level.

The element acquires the sign of $s_{\tau(i)} - s'_{\tau(i)}$ in the composition. We graphically represent element (3.7) of the first level in a rectangle, which is placed directly above the integer

$$m + n - \tau(i) + 1$$

of the ground level to accomodate this representation with the one adopted for elements of the next levels.

Similarly, depending on the sign of

$$s_{\tau(i)} - s'_{\tau(i)}$$

the rectangle lies above or below a horizontal line linked to the level. An edge between an element

$$(m + n - \tau(i) + 1)$$

of the ground level and an element

$$(s_{\tau(i)} - s'_{\tau(i)})(m + n - \tau(j) + 1)$$

of the first level is used to indicate the connection between them.

Set

$$r_{p0} = 0, \quad p = 1, ..., k$$

and

$$r_{1i} = \tau(i), i = 1, ..., N_1$$

for further use.

Elements of the set

$$WR(s, s', n, m, I_n)$$

belonging to levels

$$2 \leq p \leq k = C(s, s', n)$$

and their connections can be defined by induction as we elevate from the first level to the kth one. At each pth $p = 1, ..., k-1$ step of the induction elements of the $(p + 1)$th level and their connections with elements of the pth level are defined with the help of a procedure that partitions elements of the pth level into N_{p+1} parts.

Let κ_{pq} denote the number of the set's elements that on the pth level $p = 1, ..., k - 1$ are included by the procedure into the first q parts $q = 1, ..., N_{p+1}$.

Set

$$\kappa_{p0} = 0, \quad p = 1, .., k-1.$$

The definition of the set

$$WR(s, s', n, m, I_n)$$

results from the fact that its qth $q = 1, ..., N_{p+1}$ element of the $(p+1)$th level is the composition of elements that belong to the qth part on the pth $p = 1, ..., k-1$ level.

The set

$$WR(s, s', n, m, I_n)$$

is not defined on the $(k+1)$th level since the procedure is not able to partition all elements of the set on the kth level into parts with each of them giving the composition on the $(k+1)$th level.

In particular, we will show that the qth $q = 1, ..., N_p$ element of the set on the pth $2 \le p \le k$ level is an integer relation of the form

$$\sum_{j=r_{p,q-1}+1}^{r_{pq}} (s_{\tau(j)} - s'_{\tau(j)})(m+n-\tau(j)+1)^{p-2} = 0, \tag{3.8}$$

with

$$\sum_{j=r_{p,q-1}+1}^{r} (s_{\tau(j)} - s'_{\tau(j)})(m+n-\tau(j)+1)^{p-2} \neq 0, \tag{3.9}$$

where $r_{p,q-1} \le r < r_{pq}$ and

$$\sum_{j=r_{p,q-1}+1}^{r_{pq}} (s_{\tau(j)} - s'_{\tau(j)})(m+n-\tau(j)+1)^{i} = 0, \tag{3.10}$$

for $p \ge 3$, $i = 0, ..., p-3$.

Element (3.8) is specified by two numbers $r_{p,q-1}, r_{pq}$, where

$$r_{p0} = 0, \quad r_{p,q-1} < r_{pq}, \quad q = 1, ..., N_p, \quad r_{pN_p} \le n, \tag{3.11}$$

and N_p is the number of elements of the set on the pth level. Numbers (3.11) determine through the function τ integers

$$m+n-\tau(j)+1, \quad j = r_{p,q-1}+1, ..., r_{pq}$$

that are included in the qth element's representation (3.8).

The base case $p = 1$. The procedure makes parts and numbers them sequentially as it proceeds with elements of the first level. To have the first part the procedure, starting with the first element

$$(s_{\tau(1)} - s'_{\tau(1)})(m+n-\tau(1)+1),$$

successively adds up elements of the level to the part until for some

$$\kappa_{11}, \quad 1 < \kappa_{11} \leq N_1$$

gets an integer relation

$$\sum_{j=\kappa_{10}+1}^{\kappa_{11}} (s_{\tau(j)} - s'_{\tau(j)})(m+n-\tau(j)+1)^0 = 0 \tag{3.12}$$

as well as

$$\sum_{j=\kappa_{10}+1}^{\kappa} (s_{\tau(j)} - s'_{\tau(j)})(m+n-\tau(j)+1)^0 \neq 0, \tag{3.13}$$

where

$$\kappa_{10} + 1 \leq \kappa < \kappa_{11}.$$

The integer relation (3.12) is the first element of the set

$$WR(s, s', n, m, I_n)$$

belonging to the second level and is the composition of elements

$$(s_{\tau(j)} - s'_{\tau(j)})(m+n-\tau(j)+1), \quad j = \kappa_{10}+1, ..., \kappa_{11}$$

of the set belonging to the first level.

Set $r_{21} = \kappa_{11}$ to specify (3.12) by numbers r_{20}, r_{21}. If $\kappa_{11} = N_1$ then the procedure stops by setting $N_2 = 1$ as all elements of the set belonging to the first level are included in one part.

In this case the integer relation (3.12) is the first equation of the system (3.1)

$$\sum_{j=\kappa_{10}+1}^{\kappa_{11}} (s_{\tau(j)} - s'_{\tau(j)})(m+n-\tau(j)+1)^0$$

$$= \sum_{j=1}^{n} (s_j - s'_j)(m+n-j+1)^0 = 0,$$

confirming that all elements of the set on the first level can be put together in one part such that (3.12) holds, where

$$s_j - s'_j = 0, \quad j \neq \tau(1), ..., \tau(N_1).$$

If $\kappa_{11} < N_1$ then the procedure continues to partition the rest $N_1 - \kappa_{11}$ elements into subsequent parts. Since the number of elements of the set on the first level is finite, then there is an integer $N_2 > 1$ such that the last element on the first level is taken into consideration when the procedure makes the N_2th part. Clearly, before the rest elements on the first level are partitioned into $N_2 - 1$ parts as follows.

Let $N_2 > 2$ and suppose that q parts $1 \le q < N_2 - 1$ are defined. This means that each ith part $1 \le i \le q$ is specified by numbers $\kappa_{1,i-1}, \kappa_{1i}$ and its elements

$$(s_{\tau(j)} - s'_{\tau(j)})(m + n - \tau(j) + 1), \quad j = \kappa_{1,i-1} + 1, ..., \kappa_{1i}$$

form the composition

$$\sum_{j=\kappa_{1,i-1}+1}^{\kappa_{1i}} (s_{\tau(j)} - s'_{\tau(j)})(m + n - \tau(j) + 1)^0 = 0, \tag{3.14}$$

which is the ith element of the set on the second level. The element is specified by numbers

$$r_{2,i-1} = \kappa_{1,i-1}, r_{2i} = \kappa_{1i}.$$

We also have

$$\sum_{j=\kappa_{1,i-1}+1}^{\kappa} (s_{\tau(j)} - s'_{\tau(j)})(m + n - \tau(j) + 1)^0 \neq 0,$$

where

$$\kappa_{1,i-1} + 1 \le \kappa < \kappa_{1i}.$$

To make the next $(q+1)$th part the procedure, starting with the $(\kappa_{1q} + 1)$th element

$$(s_{\tau(\kappa_{1q}+1)} - s'_{\tau(\kappa_{1q}+1)})(m + n - \tau(\kappa_{1q} + 1) + 1),$$

successively adds up elements of the level to the $(q+1)$th part $1 \le q \le N_2 - 1$ until for some $\kappa_{1,q+1}$ gets an integer relation

$$\sum_{j=\kappa_{1q}+1}^{\kappa_{1,q+1}} (s_{\tau(j)} - s'_{\tau(j)})(m + n - \tau(j) + 1)^0 = 0, \tag{3.15}$$

as well as

$$\sum_{j=\kappa_{1q}+1}^{\kappa} (s_{\tau(j)} - s'_{\tau(j)})(m + n - \tau(j) + 1)^0 \neq 0, \tag{3.16}$$

where

$$\kappa_{1q} + 1 \le \kappa < \kappa_{1,q+1}.$$

The integer relation (3.15) is the $(q+1)$th element of the set on the second level and is the composition of the set's elements

$$(s_{\tau(j)} - s'_{\tau(j)})(m+n-\tau(j)+1), \quad j = \kappa_{1q}+1, \dots, \kappa_{1,q+1}$$

on the first level. Set $r_{2,q+1} = \kappa_{1,q+1}$ to specify (3.15) by numbers $r_{2q}, r_{2,q+1}$.

Therefore, for elements of the set on the first level not included in the first $N_2 - 1$ parts, by using the first equation of the system (3.1) and (3.14), we have

$$\sum_{j=\kappa_{1,N_2-1}+1}^{N_1} (s_{\tau(j)} - s'_{\tau(j)})(m+n-\tau(j)+1)^0$$
$$= \sum_{j=1}^{N_1} (s_{\tau(j)} - s'_{\tau(j)})(m+n-\tau(j)+1)^0$$
$$- \sum_{i=1}^{N_2-1} \sum_{j=\kappa_{1,i-1}+1}^{\kappa_{1i}} (s_{\tau(j)} - s'_{\tau(j)})(m+n-\tau(j)+1)^0 = 0, \tag{3.17}$$

where $1 \le i \le N_2 - 1$. Set $r_{1N_2} = N_1$.

Taking care of (3.17), we see that the N_2th part is made by the procedure as the last element of the first level is taken into consideration. Thus all elements of the set on the first level can be partitioned by the procedure into $N_2 \ge 1$ parts such that elements of each qth $q = 1, \dots, N_2$ part form the composition. This composition is the qth element of the set

$$WR(s, s', n, m, I_n)$$

on the second level. The base case is completed.

Inductive assumption. Suppose that the definition is made up to the pth level $2 \le p \le k-1$. In particular, conditions (3.8), (3.9) and (3.14) hold up to the pth level.

Given these assumptions, we define elements of the set on the $(p+1)$th level as the compositions of the set's elements on the pth level. The definition is made with the help of the procedure partitioning elements of the set on the pth level into N_{p+1} parts such that the composition of elements of each qth part $1 \le q \le N_{p+1}$ is the qth element of the set on the $(p+1)$th level.

To have the first part the procedure starts with the first element of the set on the pth level

$$\sum_{j=r_{p0}}^{r_{p1}} (s_{\tau(j)} - s'_{\tau(j)})(m+n-\tau(j)+1)^{p-2} = 0. \tag{3.18}$$

If

$$\sum_{j=r_{p0}}^{r_{p1}} (s_{\tau(j)} - s'_{\tau(j)})(m+n-\tau(j)+1)^{p-1} = 0, \tag{3.19}$$

then the first part is made with $\kappa_{p+1,1} = 1$.

The integer relation (3.19) is the first element of the set on the $(p+1)$th level and the composition of (3.18). Set $r_{p+1,1} = r_{p1}$ to specify (3.19) by numbers $r_{p+1,0}, r_{p+1,1}$. If in addition $N_p = 1$ then (3.19) is the $(p+1)$th equation of the system (3.1) confirming that the only element on the pth level can be put in one part such that (3.19) holds. In this case the procedure sets $N_{p+1} = 1$ and stops, otherwise it proceeds with the second part.

If (3.19) does not hold then the procedure successively adds up elements of the level to the first part until for some

$$\kappa_{p1}, \quad 1 < \kappa_{p1} \leq N_p$$

they form the composition

$$\sum_{i=\kappa_{p0}+1}^{\kappa_{p1}} \sum_{j=r_{p,i-1}+1}^{r_{pi}} (s_{\tau(j)} - s'_{\tau(j)})(m+n-\tau(j)+1)^{p-1}$$

$$= \sum_{j=r_{p+1,0}+1}^{r_{p+1,1}} (s_{\tau(j)} - s'_{\tau(j)})(m+n-\tau(j)+1)^{p-1} = 0, \tag{3.20}$$

with

$$\sum_{i=\kappa_{p0}+1}^{\kappa} \sum_{j=r_{p,i-1}+1}^{r_{pi}} (s_{\tau(j)} - s'_{\tau(j)})(m+n-\tau(j)+1)^{p-1}$$

$$= \sum_{j=r_{p+1,0}+1}^{r_{p\kappa}} (s_{\tau(j)} - s'_{\tau(j)})(m+n-\tau(j)+1)^{p-1} \neq 0, \tag{3.21}$$

where $r_{p+1,1} = r_{p,\kappa_{p1}}$ and $\kappa_{p0} + 1 \leq \kappa < \kappa_{p1}$.

Clearly, by the inductive assumption we obtain

$$\sum_{i=\kappa_{p0}+1}^{\kappa} \sum_{j=r_{p,i-1}+1}^{r_{pi}} (s_{\tau(j)} - s'_{\tau(j)})(m+n-\tau(j)+1)^{l}$$

$$= \sum_{j=r_{p+1,0}+1}^{r_{p\kappa}} (s_{\tau(j)} - s'_{\tau(j)})(m+n-\tau(j)+1)^{l} = 0, \tag{3.22}$$

for $l = 0, ..., p-2$ and

$$\kappa_{p0} + 1 \leq \kappa \leq \kappa_{p1}.$$

In particular, when $\kappa = \kappa_{p1}$ we have

$$\sum_{i=\kappa_{p0}+1}^{\kappa_{p1}} \sum_{j=r_{p,i-1}+1}^{r_{pi}} (s_{\tau(j)} - s'_{\tau(j)})(m+n-\tau(j)+1)^l$$

$$= \sum_{j=r_{p+1,0}+1}^{r_{p+1,1}} (s_{\tau(j)} - s'_{\tau(j)})(m+n-\tau(j)+1)^l = 0, \qquad (3.23)$$

where $l = 0, ..., p-2$.

The integer relation (3.20) is the first element of the set on the $(p+1)$th level and the composition of elements of the first part. It is specified by numbers $r_{p+1,0}, r_{p+1,1}$. If $\kappa_{p1} = N_p$ then (3.20) is the $(p+1)$th equation of the system (3.1) confirming that all N_p elements of the pth level can be put in one part such that (3.20) holds. In this case the procedure sets $N_{p+1} = 1$ and stops as all elements of the set on the pth level are put together into one part, otherwise it continues.

Since the number of elements of the set on the pth level is finite, then there is an integer $N_{p+1} > 1$ such that the last element of the level is taken into consideration when the procedure makes the N_{p+1}th part.

Clearly, before the rest elements of the set on the pth level are partitioned into $N_{p+1}-1$ parts as follows. Let $N_{p+1} > 2$ and suppose that q parts

$$1 \leq q < N_{p+1} - 1$$

producing compositions on the $(p+1)$th level and containing κ_{pq} elements of the pth level are defined by the procedure. Then to make the next $(q+1)$th part the procedure starts with the $(\kappa_{pq}+1)$th element

$$\sum_{j=r_{p,\kappa_{pq}}+1}^{r_{p,\kappa_{pq}+1}} (s_{\tau(j)} - s'_{\tau(j)})(m+n-\tau(j)+1)^{p-2} = 0. \qquad (3.24)$$

If

$$\sum_{j=r_{p,\kappa_{pq}}+1}^{r_{p,\kappa_{pq}+1}} (s_{\tau(j)} - s'_{\tau(j)})(m+n-\tau(j)+1)^{p-1} = 0, \qquad (3.25)$$

then the $(q+1)$th part is made with

$$\kappa_{p,q+1} = \kappa_{pq} + 1.$$

The integer relation (3.25) is the $(q+1)$th element of the set on the $(p+1)$th level and the composition of (3.24). Set

$$r_{p+1,q+1} = r_{p,\kappa_{pq}+1}$$

to specify (3.25) by numbers $r_{p+1,q+1}, r_{p+1,q}$.

If (3.25) is not true then the procedure successively adds up elements of the set on the pth level to the part until for some $\kappa_{p,q+1}$ they form the composition

$$\sum_{i=\kappa_{pq}+1}^{\kappa_{p,q+1}} \sum_{j=r_{p,i-1}+1}^{r_{pi}} (s_{\tau(j)} - s'_{\tau(j)})(m+n-\tau(j)+1)^{p-1}$$

$$= \sum_{j=r_{p+1,q}+1}^{r_{p+1,q+1}} (s_{\tau(j)} - s'_{\tau(j)})(m+n-\tau(j)+1)^{p-1} = 0, \tag{3.26}$$

with

$$\sum_{i=\kappa_{pq}+1}^{\kappa} \sum_{j=r_{p,i-1}+1}^{r_{pi}} (s_{\tau(j)} - s'_{\tau(j)})(m+n-\tau(j)+1)^{p-1}$$

$$= \sum_{j=r_{p+1,q}+1}^{r_{p\kappa}} (s_{\tau(j)} - s'_{\tau(j)})(m+n-\tau(j)+1)^{p-1} \neq 0,$$

where $r_{p+1,q+1} = r_{p\kappa_{p,q+1}}$ and

$$\kappa_{pq} + 1 \leq \kappa < \kappa_{p,q+1}.$$

By the inductive assumption we get for

$$\kappa_{pq} + 1 \leq \kappa \leq \kappa_{p,q+1}$$

and $l = 0, ..., p-2$

$$\sum_{i=\kappa_{pq}+1}^{\kappa} \sum_{j=r_{p,i-1}+1}^{r_{pi}} (s_{\tau(j)} - s'_{\tau(j)})(m+n-\tau(j)+1)^{l}$$

$$= \sum_{j=r_{p+1,q}+1}^{r_{p\kappa}} (s_{\tau(j)} - s'_{\tau(j)})(m+n-\tau(j)+1)^{l} = 0.$$

In particular, when $\kappa = \kappa_{p,q+1}$ we have for $l = 0, ..., p-2$

$$\sum_{j=r_{p+1,q}+1}^{r_{p+1,q+1}} (s_{\tau(j)} - s'_{\tau(j)})(m+n-\tau(j)+1)^{l} = 0. \tag{3.27}$$

Therefore, for elements of the set on the pth level not included in the first $(N_{p+1}-1)$ parts, by using the $(p+1)$th equation of the system (3.1) and (3.26) for $q = 0, ..., N_{p+1}-2$ with $i = q+1$, we obtain

$$\sum_{i=\kappa_{p,N_{p+1}-1}+1}^{N_p} \sum_{j=r_{p,i-1}+1}^{r_{pi}} (s_{\tau(j)} - s'_{\tau(j)})(m+n-\tau(j)+1)^{p-1}$$

$$= \sum_{j=1}^{r_{pN_p}} (s_{\tau(j)} - s'_{\tau(j)})(m+n-\tau(j)+1)^{p-1}$$

$$- \sum_{i=1}^{\kappa_{p,N_{p+1}-1}} \sum_{j=r_{p,i-1}+1}^{r_{p,i}} (s_{\tau(j)} - s'_{\tau(j)})(m+n-\tau(j)+1)^{p-1} = 0. \quad (3.28)$$

Set $\kappa_{pN_{p+1}} = N_p$. Taking care of (3.28), we see that the N_{p+1}th part is made by the procedure as the last element of the set on the pth level is taken into consideration.

Thus all elements of the set on the pth level can be partitioned by the procedure into $N_{p+1} \geq 1$ parts such that elements of each qth $q = 1, ..., N_{p+1}$ part form the composition. This composition is the qth element of the set on the $(p+1)$th level, as required.

Next, we show that this process can not be extended to the $(k+1)$th level. Indeed, by using the notation adopted, let us suppose that the procedure tries to make the N_{k+1}th part when the last element of the set on the k level is taken into consideration.

If $N_{k+1} > 1$ then for the rest elements of the set on the kth level, i.e., elements not included by the procedure in the first $N_{k+1} - 1$ parts, by using inequality (3.2), we obtain

$$\sum_{i=\kappa_{k,N_{k+1}-1}+1}^{N_k} \sum_{j=r_{k,i-1}+1}^{r_{ki}} (s_{\tau(j)} - s'_{\tau(j)})(m+n-\tau(j)+1)^{k-1}$$

$$= \sum_{j=1}^{r_{kN_k}} (s_{\tau(j)} - s'_{\tau(j)})(m+n-\tau(j)+1)^{k-1}$$

$$- \sum_{i=1}^{\kappa_{k,N_{k+1}-1}} \sum_{j=r_{k,i-1}+1}^{r_{ki}} (s_{\tau(j)} - s'_{\tau(j)})(m+n-\tau(j)+1)^{k-1} \neq 0.$$

If $N_{k+1} = 1$ then the procedure tries to put all N_k elements of the set on the kth level in one part. By using inequality (3.2), we have

$$\sum_{i=1}^{N_k} \sum_{j=r_{k,i-1}+1}^{r_{ki}} (s_{\tau(j)} - s'_{\tau(j)})(m+n-\tau(j)+1)^{k-1}$$

$$= \sum_{j=1}^{n} (s_j - s'_j)(m+n-j+1)^{k-1} \neq 0.$$

This implies that all elements of the set on the kth level can not be partitioned by the procedure into parts such that all can form the compositions on the next level.

2. ISOMORPHIC HIERARCHICAL FORMATIONS OF INTEGER PATTERNS

In the previous section we described a hierarchical set

$$WR(s, s', n, m, I_n)$$

of $k = C(s, s', n)$ levels whose elements were integers and integer relations. and proposed a notion of composition of integer relations.

Conceptually, the set $WR(s, s', n, m, I_n)$ represents a hierarchical formation of integer relations associated with the system (3.1) and inequality (3.2). The notion of composition contains the idea that integer relation can be formed from other integer relations much as molecules are formed from atoms.

Integer and integer relation are an integral part of our mental equipment but the scope of humans to fully appreciate integer relation is quite limited. Integer relation is but a set of abstract symbols subject to operational rules of arithmetic. Usually, integer relations appear as solutions to Diophantine equations and questions about their existence and character are of the main interest.

However, integer relations do not appear to us as rooted in firm reality and have the power to evoke in our minds images of concrete objects that form into each other. As a result the notion of composition of integer relations is too abstract to give us insight about what we think the formation is all about. We can not imagine how integer relations can be really formed from other integer relations, because it is not possible to visualize how this may actually happen.

In this section we extend the scope of integer relation by developing a new perception of it as a geometrical object. It is shown that integer relations have a dual existence: on one hand they are, of course, of arithmetic character, on the other hand they find an "incarnation" as two-dimensional geometrical objects that form into each other.

This translates the notion of composition and the organizing principle of integer relations into geometrical terms. Thereafter they can be visibly perceived through transformations of these geometrical objects. We show what geometrical transformation is used to get from integer patterns, corresponding to integer relations, an integer pattern corresponding to the composition of these integer relations. Importantly, this transformation can be visualized.

In particular, in this section we demonstrate that the hierarchical set of integer relations

$$WR(s, s', n, m, I_n)$$

is isomorphic to a hierarchical set of integer patterns

$$WP_{\delta\varepsilon}(s, s', n, m, I_n),$$

and moreover

$$\left\{ \begin{array}{c} \textit{Hierarchical Formation} \\ \textit{of Integer Relations} \\ WR(s, s', n, m, I_n) \end{array} \right\}$$

$$\overset{\textit{isomorphism}}{\Longleftrightarrow} \left\{ \begin{array}{c} \textit{Hierarchical Formation} \\ \textit{of Integer Patterns} \\ WP_{\delta\varepsilon}(s, s', n, m, I_n) \end{array} \right\}$$

as the set

$$WP_{\delta\varepsilon}(s, s', n, m, I_n)$$

represents a hierarchical formation of integer patterns.

Interestingly, integer relations of the set

$$WR(s, s', n, m, I_n)$$

through the isomorphism find an "incarnation" as two-dimensional geometrical objects that form into each other. In particular, they acquire shape and can be measured, for example, by their area.

The isomorphism associates the hierarchical formation of integer patterns

$$WP_{\delta\varepsilon}(s, s', n, m, I_n)$$

with the system (3.1) and inequality (3.2) and thus

$$\left\{ \begin{array}{c} \textit{System of Integer Equations (3.1)} \\ \textit{and Inequality (3.2)} \end{array} \right\}$$

$$\Downarrow$$

$$\left\{ \begin{array}{c} \textit{Hierarchical Formation} \\ \textit{of Integer Relations} \\ WR(s, s', n, m, I_n) \end{array} \right\}$$

$$\underset{\textit{isomorphism}}{\Longleftrightarrow} \left\{ \begin{array}{c} \textit{Hierarchical Formation} \\ \textit{of Integer Patterns} \\ WP_{\delta\varepsilon}(s, s', n, m, I_n) \end{array} \right\}$$

To proceed with the definition of the isomorphism we need a notion of composition of integer patterns, much as we have it for integer relations.

Definition 3.2 (Composition of Integer Patterns). *Let a function*

$$f \in W_{\delta\varepsilon}([t_m, t_{m+n}])$$

and

$$[t_{m+l_{2i-1}}, t_{m+l_{2i}}], \quad i = 1, ..., N,$$

$$0 \leq l_0, \quad l_{2i-1} < l_{2i}, \quad i = 1, ..., N,$$

$$l_{2i} < l_{2i+1}, \ i = 1, ..., N-1, \quad l_{2N-1} \leq n$$

be a set of $N \geq 1$ intervals such that patterns

$$P(f^{[k]}, [t_{m+l_{2i-1}}, t_{m+l_{2i}}]) \tag{3.29}$$

of the kth $k = 0, 1, ...$ integral of the function f are integer patterns and

$$f^{[k]}(t) = 0, \quad t \in [t_m, t_{m+n}] \setminus \bigcup_{i=1}^{N} [t_{m+l_{2i-1}}, t_{m+l_{2i}}].$$

If the pattern

$$P(f^{[k+1]}, [t_{m+l_0}, t_{m+l_{2N}}]), \tag{3.30}$$

resulting from the integration of the kth integral $f^{[k]}$ is an integer pattern, then it is called the composition of integer patterns (3.29) *and in turn* (3.29) *are called the decomposition of* (3.30).

We consider integer patterns of the $(k+1)$th integral $f^{[k+1]}$ as elements of the $(k+2)$th level $k = 0, 1...$ of a structure and represent the composition of integer patterns with a help of a diagram. In the diagram the $(k+2)$th level has a coordinate system with the graph of the $(k+1)$th integral $f^{[k+1]}$.

This coordinate system is placed at some eye-pleasing distance above the $(k+1)$th level which has a coordinate system with the graph of the kth integral $f^{[k]}$ (see Figure 29 as an illustration). In the diagram integer patterns (3.29) on the $(k+1)$th level are connected with the integer pattern (3.30) on the $(k+2)$th level by edges to show the connection between them in the composition.

Thus in the diagram the composition of integer patterns takes the form of a tree on the $(k+1)$th and $(k+2)$th levels with vertices being the integer patterns. In diagrammatic language the tree reads that the integration of the kth integral $f^{[k]}$ results in the composition of integer patterns (3.29) on the $(k+1)$th level and this leads to the formation of the integer pattern (3.30) on the $(k+2)$th level.

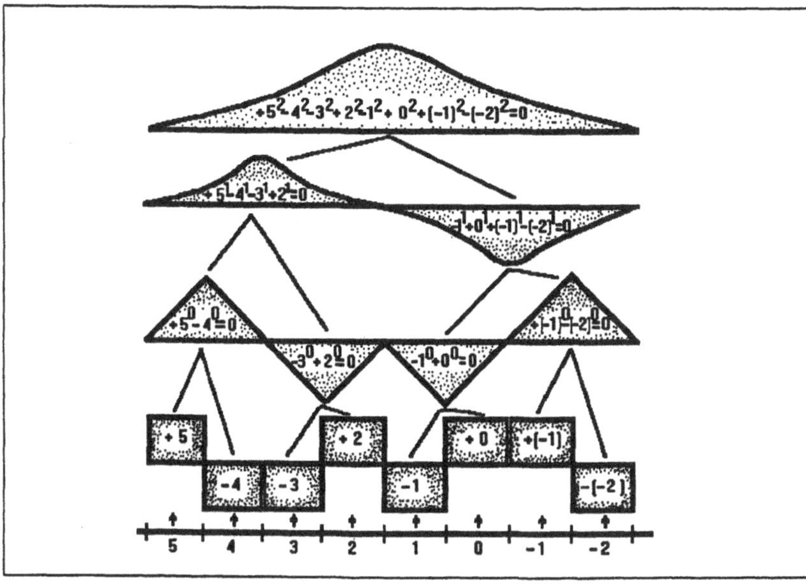

Figure 29: Two hierarchical formations of integer patterns and integer relations are displayed together to underline the unified character of elements of the web of relations. The figure actually allows us to see how integer relations form into each other. Note that integer relations can be measured by the area of corresponding integer patterns.

Clearly, the tree can be read in the reverse order, i.e., the differentiation of the $(k+1)$th integral $f^{[k+1]}$ results in the decomposition of the integer pattern (3.30) on the $(k+2)$th level and gives integer patterns (3.29) on the $(k+1)$th level. Therefore, the composition of integer patterns has an absolute nature as it can be visualized through the integration of functions.

Next, we develop a new perception of integer relations and their transformations. Namely, we define an isomorphism $\psi_{\delta\varepsilon}$ between elements of the set

$$WR(s, s', n, m, I_n),$$

i.e., integers and integer relations, and integer patterns of integrals $f^{[p]}, p = 0, \ldots, k-1$ of a function

$$f = \rho_{m\delta\varepsilon}(s - s') \in W_{\delta\varepsilon}[t_m, t_{m+n}]$$

with the composition preserved under the map $\psi_{\delta\varepsilon}$.

As a result we define a set of integer patterns

$$WP_{\delta\varepsilon}(s, s', n, m, I_n)$$

which is isomorphic to the set of integer relations

$$WR(s, s', n, m, I_n),$$

i.e.,

$$\psi_{\delta\varepsilon} : WR(s, s', n, m, I_n) \Longleftrightarrow WP_{\delta\varepsilon}(s, s', n, m, I_n).$$

The definition is done by induction as we elevate from the first level of the set $WR(s, s', n, m, I_n)$ to the kth one. The formula (2.1) of the integer code series and the geometrical interpretation of structural numbers are our main working instrument in the subsequent expressions for

$$f^{[p]}(t_j), \quad p = 1, 2, \ldots, \quad j = m+1, \ldots, m+n$$

where, to simplify notation we suppress ε and δ dependence by letting $\varepsilon = 1$ and $\delta = 1$. It will be resumed after the definition of the isomorphism.

Firstly, we define the map $\psi_{\delta\varepsilon}$ between elements of the set

$$WR(s, s', n, m, I_n)$$

on the first level and all integer patterns of the function f, i.e.,

$$\psi_{\delta\varepsilon} : \{\textit{elements of } WR(s, s', n, m, I_n) \textit{ on the first level}\}$$

$$\Updownarrow$$

$$\{\textit{integer patterns of } f\},$$

or more explicitly by

$$\psi_{\delta\varepsilon} : (s_{\tau(i)} - s'_{\tau(i)})(m + n - \tau(i) + 1)$$

$$\Updownarrow$$

$$P(f, [t_{m+\tau(i)-1}, t_{m+\tau(i)}]),$$

where $i = 1, ..., N_1$. Clearly, the map is one-to-one on the first levels.

The base case $p = 1$. We partition the integer patterns on the first level into N_2 parts to have the following properties. The partition is preserved under the map $\psi_{\delta\varepsilon}$ which can be extended to the second level so that the composition of integer patterns of the qth part $q = 1, ..., N_2$ is in a one-to-one correspondence with the composition of integers included into the qth part by the procedure.

We prove the statement by the second induction on the number of the parts q. For $q = 1$ and $j = 1, ..., \kappa_{11}$ consider integer patterns

$$P(f, [t_{m+\tau(j)-1}, t_{m+\tau(j)}]), \tag{3.31}$$

which clearly correspond to integers of the first part and the pattern

$$P(f^{[1]}, [t_{m+\tau(\kappa_{10}+1)-1}, t_{m+\tau(\kappa_{11})}]). \tag{3.32}$$

We wish to prove that the pattern (3.32) is an integer pattern and thus the composition of integer patterns (3.31). This needs to show that

$$f^{[1]}(t_{m+\tau(\kappa_{10}+1)-1}) = 0, \quad f^{[1]}(t_{m+\tau(\kappa_{11})}) = 0,$$

and

$$f^{[1]}(t_{m+\tau(j)}) \neq 0, \tag{3.33}$$

when $\kappa_{10} + 1 \leq j < \kappa_{11}$.

If $\tau(\kappa_{10} + 1) = 1$ then $f^{[1]}(t_m) = 0$ by definition. If $\tau(\kappa_{10} + 1) > 1$ then

$$f^{[1]}(t_{m+\tau(\kappa_{10}+1)-1})$$

$$= \alpha_{1m0} \sum_{j=1}^{\tau(\kappa_{10}+1)-1} (s_j - s'_j)(m + \tau(\kappa_{10} + 1) - j + 1)^0 = 0,$$

since $s_j - s'_j = 0$ for $j = 1, ..., \tau(\kappa_{10} + 1) - 1$.

We have

$$f^{[1]}(t_{m+\tau(\kappa_{11})}) = \alpha_{1m0} \sum_{j=1}^{\tau(\kappa_{11})} (s_j - s'_j)(m + \tau(\kappa_{11}) - j + 1)^0$$

$$= \alpha_{1m0} \sum_{j=\kappa_{10}+1}^{\kappa_{11}} (s_{\tau(j)} - s'_{\tau(j)})(m + n - \tau(j) + 1)^0 = 0, \tag{3.34}$$

since for all $j = 1, ..., \tau(\kappa_{11})$, other than $j = \tau(\kappa_{10}+1), ..., \tau(\kappa_{11})$, the condition $s_j - s'_j = 0$ holds. From (3.34) we obtain immediately that

$$f^{[1]}(t_{m+\tau(\kappa_{11})}) = 0. \tag{3.35}$$

Note how the first element of the set

$$WR(s, s', n, m, I_n)$$

on the second level and $f^{[1]}(t_{m+\tau(\kappa_{11})})$ are connected in (3.34), and that this completely determines (3.35).

If (3.33) is not true, then clearly for

$$\kappa_{10} + 1 \leq \kappa < \kappa_{11}$$

we get

$$\sum_{j=\kappa_{10}+1}^{\kappa} (s_{\tau(j)} - s'_{\tau(j)})(m + n - \tau(j) + 1)^0 = 0,$$

contradicting (3.33).

Thus the pattern (3.32) is an integer pattern and the composition of the integer patterns (3.35). It is defined to be the first element of the set

$$WP_{\delta\varepsilon}(s, s', n, m, I_n)$$

on the second level.

Then by sending

$$\sum_{j=\kappa_{10}+1}^{\kappa_{11}} (s_{\tau(j)} - s'_{\tau(j)})(m + n - \tau(j) + 1)^0 = 0$$

$$\Updownarrow$$

$$P(f^{[1]}, [t_{m+\tau(\kappa_{10}+1)-1}, t_{m+\tau(\kappa_{11})}])$$

the map $\psi_{\delta\varepsilon}$ can be extended to preserve the composition. Namely, the composition of integers of the first part turns out to be related with the composition of corresponding integer patterns of the first part.

If $\kappa_{11} = N_1$ we are done with the base case, otherwise we continue with the second induction. Suppose that the isomorphism $\psi_{\delta\varepsilon}$ is extended up to q parts $1 \leq q \leq N_2 - 1$ on the first level and their compositions on the second level, we establish it for the $(q+1)$th part.

We put into the $(q+1)$th part integer patterns

$$P(f, [t_{m+\tau(j)-1}, t_{m+\tau(j)}]), \tag{3.36}$$

where $j = \kappa_{1q}+1, ..., \kappa_{1,q+1}$. These integer patterns clearly correspond to the integers included by the procedure into the $(q+1)$th part.

Similarly, we wish to show that the pattern

$$P(f^{[1]}, [t_{m+\tau(\kappa_{1q}+1)-1}, t_{m+\tau(\kappa_{1,q+1})}]) \tag{3.37}$$

is an integer one and the composition of (3.36) by proving that

$$f^{[1]}(t_{m+\tau(\kappa_{1q}+1)-1}) = 0, \quad f^{[1]}(t_{m+\tau(\kappa_{1,q+1})}) = 0$$

and

$$f^{[1]}(t_{m+\tau(j)}) \neq 0, \tag{3.38}$$

where $\kappa_{1q}+1 \leq j < \kappa_{1,q+1}$.

By using properties of the first $(q+1)$ parts, it is easy to check that

$$f^{[1]}(t_{m+\tau(\kappa_{1q}+1)-1})$$

$$= \alpha_{1m0} \sum_{j=1}^{\tau(\kappa_{1q}+1)-1} (s_j - s'_j)(m+\tau(\kappa_{1q}+1)-1-j+1)^0$$

$$= \alpha_{1m0} \sum_{j=1}^{\tau(\kappa_{1q}+1)-1} (s_j - s'_j)(m+n-j+1)^0$$

$$= \alpha_{1m0} \sum_{i=1}^{q} \sum_{j=\kappa_{1,i-1}+1}^{\kappa_{1,i}} (s_{\tau(j)} - s'_{\tau(j)})(m+n-\tau(j)+1)^0 = 0, \tag{3.39}$$

since $\tau(\kappa_{1q}+1)-1 = \tau(\kappa_{1q})$ if

$$s_{\tau(\kappa_{1q})+1} - s'_{\tau(\kappa_{1q})+1} \neq 0$$

and $s_j - s'_j = 0$ for

$$\tau(\kappa_{1q})+1 \leq j \leq \tau(\kappa_{1q}+1)-1$$

otherwise.

By applying (3.39), we have

$$f^{[1]}(t_{m+\tau(\kappa_{1,q+1})})$$

$$= \alpha_{1m0} \sum_{j=1}^{\tau(\kappa_{1,q+1})} (s_j - s'_j)(m+\tau(\kappa_{1,q+1})-j+1)^0$$

$$= f^{[1]}(t_{m+\tau(\kappa_{1q}+1)-1})$$

$$+\alpha_{1m0}\sum_{j=\kappa_{1,q}+1}^{\kappa_{1,q+1}}(s_{\tau(j)}-s'_{\tau(j)})(m+n-\tau(j)+1)^0=0. \qquad (3.40)$$

Note that given (3.39) the $(q+1)$th element of the set

$$WR(s,s',n,m,I_n)$$

on the second level appears explicitly in (3.40) and completely specifies it.

If (3.38) is not true then it follows immediately that for

$$\kappa_{1q}\le\kappa<\kappa_{1,q+1}$$

we obtain

$$\sum_{j=\kappa_{1q}+1}^{\kappa}(s_{\tau(j)}-s'_{\tau(j)})(m+n-\tau(j)+1)^0=0,$$

contradicting (3.16).

Thus pattern (3.37) is an integer pattern and the composition of integer patterns (3.36). It is defined to be the $(q+1)$th element of the set

$$WP_{\delta\varepsilon}(s,s',n,m,I_n)$$

on the second level.

Then by sending

$$\sum_{j=\kappa_{1q}+1}^{\kappa_{1,q+1}}(s_{\tau(j)}-s'_{\tau(j)})(m+n-\tau(j)+1)^0=0$$

$$\Updownarrow$$

$$P(f^{[1]},[t_{m+\tau(\kappa_{1q}+1)-1},t_{m+\tau(\kappa_{1,q+1})}])$$

the map $\psi_{\delta\varepsilon}$ can be extended to preserve the composition. Namely, the composition of integers of the $(q+1)$th part is defined to be related with the composition of the corresponding integer patterns of the $(q+1)$th part.

This completes the second induction on the number of the parts on the second level and the isomorphism $\psi_{\delta\varepsilon}$ is defined up to the second level.

We continue with the first induction on level p by making the assumption. Suppose that the isomorphism $\psi_{\delta\varepsilon}$ is extended up to the pth $2\le p\le k-1$ level, we establish it for the $(p+1)$th level, i.e.,

$$\psi_{\delta\varepsilon}:\{\textit{elements of } WR(s,s',n,m,I_n) \textit{ on the } (p+1)\textit{th level}\}$$

$$\Updownarrow$$

$$\{\textit{integer patterns of } f^{[p]}\}.$$

Now we partition integer patterns on the pth level into N_{p+1} parts to have the following properties. The partition is preserved under the map $\psi_{\delta\varepsilon}$, which can be extended to the $(p+1)$th level so that the composition of integer patterns of the qth part $q = 1, ..., N_{p+1}$ is in a one-to-one correspondence with the composition of integer relations included into the qth part by the procedure.

We prove the statement by starting the second induction on the number q of the parts on the pth level. For simplicity of notation, let $h_{pq} = \tau(r_{p\kappa_{pq}})$ denote the maximum component's index of sequences s, s' included in representation (3.8) of the first κ_{pq} elements on the pth level and $h^+_{pq} = \tau(r_{p\kappa_{pq}} + 1)$.

Clearly, we can set

$$h_{p0} = 0, \quad p = 1, ..., k-1.$$

For $q = 1$ and $i = \kappa_{p0} + 1, ..., \kappa_{p1}$ take integer patterns

$$P(f^{[p-1]}, [t_{m+\tau(r_{p,i-1}+1)-1}, t_{m+\tau(r_{pi})}]), \tag{3.41}$$

which clearly correspond to the integer relations of the first part on the pth level and consider the pattern

$$P(f^{[p]}, [t_{m+h^+_{p0}-1}, t_{m+h_{p1}}]). \tag{3.42}$$

We wish to prove that the pattern (3.42) is an integer pattern and thus the composition of (3.41) by first showing that

$$f^{[p]}(t_{m+h^+_{p0}-1}) = 0, \quad f^{[p]}(t_{m+h_{p1}}]) = 0.$$

If $h^+_{p0} = 1$ then by definition $f^{[p]}(t_m) = 0$. If $h^+_{p0} > 1$ then

$$f^{[p]}(t_{m+h^+_{p0}-1})$$

$$= \sum_{i=0}^{p-1} \alpha_{pmi} \sum_{j=1}^{h^+_{p0}-1} (s_j - s'_j)(m + h^+_{p0} - 1 - j + 1)^0 = 0,$$

since

$$s_j - s'_j = 0, \quad j = 1, ..., h^+_{p0} - 1.$$

Now, recalling (3.20), we obtain

$$f^{[p]}(t_{m+h_{p1}})$$

$$= \sum_{i=0}^{p-1} \alpha_{pmi} \sum_{i'=\kappa_{p0}+1}^{\kappa_{p1}} \sum_{j=r_{p,i'-1}+1}^{r_{pi'}} (s_{\tau(j)} - s'_{\tau(j)})(m + h_{p1} - \tau(j) + 1)^i$$

$$= \sum_{i=0}^{p-1} \alpha_{pmi} \sum_{j=r_{p+1,0}+1}^{r_{p+1,1}} (s_{\tau(j)} - s'_{\tau(j)})(m + h_{p1} - \tau(j) + 1)^i$$

$$= \sum_{i=0}^{p-1} \alpha_{pmi} \sum_{j=r_{p+1,0}+1}^{r_{p+1,1}} (s_{\tau(j)} - s'_{\tau(j)})((m + n - \tau(j) + 1) + (h_{p1} - n))^i$$

$$= \sum_{i=0}^{p-1} \alpha_{pmi} \sum_{j=r_{p+1,0}+1}^{r_{p+1,1}} (s_{\tau(j)} - s'_{\tau(j)}) \sum_{l=0}^{i} \binom{i}{l} (m + n - \tau(j) + 1)^l (h_{p1} - n)^{i-l}$$

$$= \sum_{i=0}^{p-1} \alpha_{pmi} \sum_{l=0}^{i} \binom{i}{l} (h_{p1} - n)^{i-l} \sum_{j=r_{p+1,0}+1}^{r_{p+1,1}} (s_{\tau(j)} - s'_{\tau(j)})(m+n-\tau(j)+1)^l. \tag{3.43}$$

Separating the term with the $(p-1)$th power of integers

$$m + n - \tau(j) + 1, \quad j = r_{p+1,0} + 1, \ldots, r_{p+1,1},$$

we get

$$f^{[p]}(t_{m+h_{p1}})$$

$$= \sum_{i=0}^{p-2} \alpha_{pmi} \sum_{l=0}^{i} \binom{i}{l} (h_{p1} - n)^{i-l} \sum_{j=r_{p+1,0}+1}^{r_{p+1,1}} (s_{\tau(j)} - s'_{\tau(j)})(m + n - \tau(j) + 1)^l$$

$$+\alpha_{pm,p-1} \sum_{l=0}^{p-2} \binom{p-1}{l} (h_{p1} - n)^{p-1-l}$$

$$\sum_{j=r_{p+1,0}+1}^{r_{p+1,1}} (s_{\tau(j)} - s'_{\tau(j)})(m + n - \tau(j) + 1)^l$$

$$+\alpha_{pm,p-1} \sum_{j=r_{p+1,0}+1}^{r_{p+1,1}} (s_{\tau(j)} - s'_{\tau(j)})(m + n - \tau(j) + 1)^{p-1}. \tag{3.44}$$

Then, by using (3.23), we get for (3.44)

$$f^{[p]}(t_{m+h_{p1}})$$

$$= \alpha_{pm,p-1} \sum_{j=r_{p+1,0}+1}^{r_{p+1,1}} (s_{\tau(j)} - s'_{\tau(j)})(m+n-\tau(j)+1)^{p-1} = 0. \quad (3.45)$$

Therefore, from (3.44) we see that the first element of the set

$$WR(s, s', n, m, I_n)$$

on the $(p+1)$th level completely specifies the condition

$$f^{[p]}(t_{m+h_{p1}}) = 0.$$

Now, we have

$$f^{[p]}(t_{m+h^+_{p0}-1}) = 0, \quad f^{[p]}(t_{m+h_{p1}}) = 0,$$

and for $i = 1, ..., p-1$

$$f^{[i]}(t_{m+h^+_{p0}-1}) = 0, \quad f^{[i]}(t_{m+h_{p1}}) = 0.$$

Points $t_{m+h^+_{p0}}, t_{m+h_{p1}}$ are candidates as endpoints of integer pattern of the pth integral $f^{[p]}$. To prove that the pattern (3.42) is an integer pattern we have to show that there are no points

$$t_{m+j}, \quad h^+_{p0} \leq j < h_{p1}$$

in the interval

$$(t_{m+h^+_{p0}-1}, t_{m+h_{p1}})$$

such that

$$f^{[i]}(t_{m+j}) = 0$$

for all $i = 1, ..., p$ at once.

By the inductive assumption the only points

$$t_{m+j}, \quad h^+_{p0} \leq j < h_{p1}$$

satisfying

$$f^{[i]}(t_{m+j}) = 0, \quad i = 1, ..., p-1$$

are endpoints

$$t_{m+j}, \quad j = \tau(r_{p,\kappa-1}+1)-1, \tau(r_{p\kappa}), \quad \kappa = \kappa_{p0}+1, ..., \kappa_{p1}, \quad (3.46)$$

of integer patterns (3.41), where points

$$t_{m+h^+_{p0}-1}, \quad t_{m+h_{p1}}$$

are excluded from the consideration.

Thus points (3.46) are the only candidates for endpoints of integer pattern of the pth integral $f^{[p]}$ in the interval. If $\kappa_{p1} = 1$ then there is no points (3.46) in the interval. If $\kappa_{p1} > 1$ then we claim that the pth integral $f^{[p]}$ does not take zero at (3.46).

Firstly, for this purpose we wish to evaluate

$$f^{[p]}(t_{m+\tau(r_{p\kappa})})$$

$$= \sum_{i=0}^{p-1} \alpha_{pmi} \sum_{i'=\kappa_{p0}+1}^{\kappa} \sum_{j=r_{p,i'-1}+1}^{r_{pi'}} (s_{\tau(j)} - s'_{\tau(j)})(m + \tau(r_{p\kappa}) - \tau(j) + 1)^i$$

$$= \sum_{i=0}^{p-1} \alpha_{pmi} \sum_{j=r_{p+1,0}+1}^{r_{p\kappa}} (s_{\tau(j)} - s'_{\tau(j)})(m + \tau(r_{p\kappa}) - \tau(j) + 1)^i.$$

Similarly to (3.43), replacing $r_{p+1,1}$ by $r_{p\kappa}$ and h_{p1} by $\tau(r_{p\kappa})$, we get

$$f^{[p]}(t_{m+\tau(r_{p\kappa})})$$

$$= \sum_{i=0}^{p-2} \alpha_{pmi} \sum_{l=0}^{i} \binom{i}{l} (\tau(r_{p\kappa}) - n)^{i-l}$$

$$\sum_{j=r_{p+1,0}+1}^{r_{p\kappa}} (s_{\tau(j)} - s'_{\tau(j)})(m + n - \tau(j) + 1)^l$$

$$+\alpha_{pm,p-1} \sum_{l=0}^{p-2} \binom{p-1}{l} (\tau(r_{p\kappa}) - n)^{p-1-l}$$

$$\sum_{j=r_{p+1,0}+1}^{r_{p\kappa}} (s_{\tau(j)} - s'_{\tau(j)})(m + n - \tau(j) + 1)^l$$

$$+\alpha_{pm,p-1} \sum_{j=r_{p+1,0}+1}^{r_{p\kappa}} (s_{\tau(j)} - s'_{\tau(j)})(m + n - \tau(j) + 1)^{p-1}. \tag{3.47}$$

Using (3.22) and (3.21), we infer from (3.47)

$$f^{[p]}(t_{m+\tau(r_{p\kappa})})$$

$$= \alpha_{pm,p-1} \sum_{j=r_{p+1,0}+1}^{r_{p\kappa}} (s_{\tau(j)} - s'_{\tau(j)})(m + n - \tau(j) + 1)^{p-1} \neq 0. \tag{3.48}$$

Secondly, by using (3.48) for

$$\kappa_{p0} + 1 \leq \kappa < \kappa_{p1}$$

we get if

$$\tau(r_{p\kappa}) + 1 = \tau(r_{p\kappa} + 1)$$

then

$$f^{[p]}(t_{m+\tau(r_{p,\kappa+1-1}+1)-1}) = f^{[p]}(t_{m+\tau(r_{p\kappa})}) \neq 0;$$

and

$$f^{[p]}(t_{m+\tau(r_{p\kappa}+1)-1}) = f^{[p]}(t_{m+\tau(r_{p\kappa})})$$

$$+\sum_{i=0}^{p-1} \alpha_{pmi} \sum_{j=\tau(r_{p\kappa})+1}^{\tau(r_{p\kappa}+1)-1} (s_j - s'_j)(m + \tau(r_{p\kappa}+1) - 1 - j + 1)^i$$

$$= f^{[p]}(t_{m+\tau(r_{p\kappa})}) \neq 0,$$

since

$$s_j - s'_j = 0, \quad j = \tau(r_{p\kappa}) + 1, ..., \tau(r_{p\kappa}+1) - 1,$$

otherwise. This proves the claim.

Thus the pattern (3.42) is an integer pattern and is the composition of integer patterns (3.41). It is defined to be the first element of the set

$$WP_{\delta\varepsilon}(s, s', n, m, I_n)$$

on the $(p+1)$th level.

Then by sending

$$\sum_{j=r_{p+1,0}+1}^{r_{p+1,1}} (s_{\tau(j)} - s'_{\tau(j)})(m + n - \tau(j) + 1)^{p-1} = 0$$

$$\Updownarrow$$

$$P(f^{[p]}, [t_{m+h^+_{p0}}, t_{m+h_{p1}}])$$

the map $\psi_{\delta\varepsilon}$ can be extended to preserve the composition. Namely, the composition of integer relations of the first part turns out to be related with the composition of corresponding integer patterns of the first part.

If $\kappa_{p1} = N_p$ we are done, otherwise we continue with the second induction. Suppose that the isomorphism $\psi_{\delta\varepsilon}$ is extended up to q parts $1 \leq q \leq N_{p+1} - 1$ on the pth level and their compositions on the $(p+1)$th level, we establish it for the $(q+1)$th part.

For $i = \kappa_{pq}+1, ..., \kappa_{p,q+1}$ we put into the $(q+1)$th part integer patterns

$$P(f^{[p-1]}, [t_{m+\tau(r_{p,i-1}+1)-1}, t_{m+\tau(r_{pi})}]), \tag{3.49}$$

which correspond to the integer relations of the $(q+1)$th part made by the procedure on the pth level and consider the pattern

$$P(f^{[p]}, [t_{m+h_{pq}^{+}-1}, t_{m+h_{p,q+1}}]). \tag{3.50}$$

Similarly, we wish to prove that the pattern (3.50) is an integer pattern and thus the composition of integer patterns (3.49) by firstly showing that

$$f^{[p]}(t_{m+h_{pq}^{+}-1}) = 0, \quad f^{[p]}(t_{m+h_{p,q+1}}]) = 0.$$

By the inductive assumption if

$$\tau(r_{p+1,q}) + 1 = \tau(r_{p+1,q}+1)$$

then

$$f^{[p]}(t_{m+h_{pq}^{+}-1}) = f^{[p]}(t_{m+\tau(r_{p+1,q}+1)-1})$$
$$= f^{[p]}(t_{m+\tau(r_{p+1,q})}) = 0$$

and

$$f^{[p]}(t_{m+\tau(r_{p+1,q}+1)-1}) = f^{[p]}(t_{m+\tau(r_{p+1,q})})$$

$$+\sum_{i=0}^{p-1} \alpha_{pmi} \sum_{j=\tau(r_{p+1,q})+1}^{\tau(r_{p+1,q}+1)-1} (s_j - s'_j)(m+\tau(r_{pq}+1)-1-j+1)^i = 0,$$

since

$$s_j - s'_j = 0, \quad j = \tau(r_{p+1,q})+1, ..., \tau(r_{p+1,q}+1)-1,$$

otherwise.

Next, we have

$$f^{[p]}(t_{m+h_{p,q+1}})$$

$$= \sum_{i=0}^{p-1} \alpha_{pmi} \sum_{i'=\kappa_{p0}+1}^{\kappa_{p,q+1}} \sum_{j=r_{p,i'-1}+1}^{r_{pi'}} (s_{\tau(j)} - s'_{\tau(j)})(m+h_{p,q+1}-\tau(j)+1)^i$$

$$= \sum_{i=0}^{p-1} \alpha_{pmi} \sum_{j=r_{p+1,0}+1}^{r_{p+1,q+1}} (s_{\tau(j)} - s'_{\tau(j)})(m+h_{p,q+1}-\tau(j)+1)^i.$$

Similarly to (3.43), replacing h_{p1} by $h_{p,q+1}$ and $r_{p+1,1}$ by $r_{p+1,q+1}$, we get

$$f^{[p]}(t_{m+h_{p,q+1}}))$$

$$= \sum_{i=0}^{p-1} \alpha_{pmi} \sum_{l=0}^{i} \binom{i}{l} (h_{p,q+1}-n)^{i-l}$$

$$\sum_{j=r_{p+1,0}+1}^{r_{p+1,q+1}} (s_{\tau(j)} - s'_{\tau(j)})(m+n-\tau(j)+1)^l.$$

Then, by separating terms connected with the first q parts and the $(p-1)$th power of integers

$$(m+n-\tau(j)+1), \quad j = r_{p+1,q}+1, ..., r_{p+1,q+1},$$

we obtain

$$f^{[p]}(t_{m+h_{p,q+1}})$$

$$= \sum_{i=0}^{p-1} \alpha_{pmi} \sum_{l=0}^{i} \binom{i}{l} (h_{p,q+1}-n)^{i-l}$$

$$\sum_{j=1}^{q} \sum_{j'=r_{p+1,j-1}+1}^{r_{p+1,j}} (s_{\tau(j')} - s'_{\tau(j')})(m+n-\tau(j')+1)^l$$

$$+ \sum_{i=0}^{p-2} \alpha_{pmi} \sum_{l=0}^{i} \binom{i}{l} (h_{p,q+1}-n)^{i-l}$$

$$\sum_{j=r_{p+1,q}+1}^{r_{p+1,q+1}} (s_{\tau(j)} - s'_{\tau(j)})(m+n-\tau(j)+1)^l$$

$$+\alpha_{pm,p-1} \sum_{l=0}^{p-2} \binom{p-1}{l} (h_{p,q+1}-n)^{p-1-l}$$

$$\sum_{j=r_{p+1,q}+1}^{r_{p+1,q+1}} (s_{\tau(j)} - s'_{\tau(j)})(m+n-\tau(j)+1)^l$$

$$+\alpha_{pm,p-1} \sum_{j=r_{p+1,q}+1}^{r_{p+1,q+1}} (s_{\tau(j)} - s'_{\tau(j)})(m+n-\tau(j)+1)^{p-1}. \tag{3.51}$$

By using (3.8) and (3.10) for the first q elements of the set

$$WR(s, s', n, m, I_n)$$

on the $(p+1)$th level and (3.27) for the $(q+1)$th part on the pth level, we get

$$f^{[p]}(t_{m+h_{p,q+1}})$$

$$= \alpha_{pm,p-1} \sum_{j=r_{p+1,q}+1}^{r_{p+1,q+1}} (s_{\tau(j)} - s'_{\tau(j)})(m+n-\tau(j)+1)^{p-1}. \tag{3.52}$$

This gives

$$f^{[p]}(t_{m+h_{p,q+1}}) = 0,$$

by taking care of (3.26).

We see from (3.51) and (3.52) that under the condition that the isomorphism is extended up to the qth part on the pth level

$$f^{[p]}(t_{m+h_{p,q+1}})$$

is completely specified by the $(q+1)$th element of the set

$$WR(s, s', n, m, I_n)$$

on the $(p+1)$th level.

Thus we have

$$f^{[p]}(t_{m+h^{+}_{pq}-1}) = 0, \quad f^{[p]}(t_{m+h_{p,q+1}}) = 0$$

and by the inductive assumption

$$f^{[i]}(t_{m+h^{+}_{pq}-1}) = 0, \quad f^{[i]}(t_{m+h_{p,q+1}}) = 0,$$

where $i = 1, ..., p-1$. Points

$$t_{m+h^{+}_{pq}-1}, t_{m+h_{p,q+1}}$$

are candidates for endpoints of integer pattern of the pth integral $f^{[p]}$.

To prove that (3.50) is an integer pattern we have to show that there are no points

$$t_{m+j}, \quad h^{+}_{pq} \leq j < h_{p,q+1}$$

in the interval

$$(t_{m+h^{+}_{pq}-1}, t_{m+h_{p,q+1}})$$

such that

$$f^{[i]}(t_{m+j}) = 0$$

for all $i = 1, ..., p$ at once. The proof is identical in form to that given for the interval $(t_{m+h^{+}_{p0}-1}, t_{m+h_{p1}})$ and we will not go through it again.

Thus the pattern (3.50) is an integer pattern and the composition of the integer patterns (3.49). It is defined to be the $(q+1)$th element of the set

$$WP_{\delta\varepsilon}(s, s', n, m, I_n)$$

on the $(p+1)$th level.

Then by sending

$$\sum_{j=r_{p+1,q}+1}^{r_{p+1,q+1}} (s_{\tau(j)} - s'_{\tau(j)})(m+n-\tau(j)+1)^{p-1} = 0$$

$$\Updownarrow$$

$$P(f^{[p]}, [t_{m+h^+_{pq}}, t_{m+h_{p,q+1}}])$$

the map $\psi_{\delta\varepsilon}$ can be extended to preserve the composition. Namely, the composition of integer relations of the $(q+1)$th part on the pth level is defined to be related with the composition of corresponding integer patterns on the pth level.

This completes the second induction on the number of the parts q on the pth level and finally the first induction on the number of levels p.

It can be easily shown that the isomorphism $\psi_{\delta\varepsilon}$ can not be extended to the $(k+1)$th level since not all integer patterns of the set

$$WP_{\delta\varepsilon}(s, s', n, m, I_n)$$

on the kth level give compositions on the $(k+1)$th level.

Indeed, if $N_{k+1} > 1$ then the isomorphism can be extended, in the manner used for the previous levels, to include on the $(k+1)$th level the compositions of the first $N_{k+1}-1$ parts on the kth levels. However, it is clear that the rest of integer patterns on the kth level do not form the composition on the $(k+1)$th level because of inequality (3.2).

If $N_{k+1} = 1$ then the same reasoning can be applied to the first part by using again (3.2).

As the result a hierarchical formation of integer patterns

$$WP_{\delta\varepsilon}(s, s', n, m, I_n)$$

consisting of $C(s, s', n)$ levels and isomorphic to the hierarchical formation of integer relations

$$WR(s, s', n, m, I_n)$$

is defined by

$$\psi_{\delta\varepsilon} : WR(s, s', n, m, I_n) \Longleftrightarrow WP_{\delta\varepsilon}(s, s', n, m, I_n).$$

Therefore, based on the system (3.1) and inequality (3.2) two isomorphic hierarchical formations of integer relations and integer patterns

$$WR(s, s', n, m, I_n), \quad WP_{\delta\varepsilon}(s, s', n, m, I_n)$$

are defined. These hierarchical formations are the main ingredient in the definition of the structure.

Giving leaving credit to the system (3.1) for its leading role in the definition of the hierarchical formations, we however should note that viewed by itself the system does not display much evidence to conclude that it is related with a hierarchical formation of two-dimensional geometrical patterns.

Without the geometrical context, it looks like a curious result that expresses a property of integers. It seems reasonable to suggest that formal methods have limited applications in deriving properties of the hierarchical formations such as the nonlocal order and large symmetry observed in section 6 of Chapter 2, because algebraic manipulations of symbols can not generate their visible images.

In particular, the system (3.1) can be easily brought into a form that is well-known in many areas. For simplicity, let

$$s = s_1 ... s_n, s' = s'_1 ... s'_n \in B_n, \quad s \neq s'$$

then $s_i - s'_i, i = 1, ..., n$ takes values $-2, 0, 2$. Similarly to remark 2.2, powers

$$(m + n + 1 - i)^j, \quad j = 0, ..., k - 2$$

of integers

$$m + n + 1 - i, \quad i = 1, ..., n$$

are included in (3.1) with $+$ sign if $s_i - s'_i = +2$, with $-$ sign if $s_i - s'_i = -2$, and are not included if

$$s_i - s'_i = 0, \quad i = 1, ..., n.$$

By using the notation in (2.39) and cancelling 2 out for convenience, the system (3.1) can be turned into

$$x_1^1 + ... + x_p^1 = y_1^1 + ... + y_p^1$$

$$. \quad . \quad . \quad .$$

$$x_1^{k-2} + ... + x_p^{k-2} = y_1^{k-2} + ... + y_p^{k-2}, \tag{3.53}$$

where in addition it is supposed that

$$C(s, s', n) = k \geq 3$$

and

$$p = q, \ 1 < p < \lfloor \frac{n}{2} \rfloor.$$

The system (3.53) showed up in many different contexts, under many guises (for example, [50], [51], [52], [53]). However, as known to the author, there are no geometrical interpretations of (3.53) leading to the connection with hierarchical formations of two-dimensional geometrical patterns that, when visualized, exhibit nonlocal order and large symmetry. In this respect the geometrical interpretation of structural numbers is indeed of crucial importance to the understanding of the system (3.1).

3. DEFINITION OF THE STRUCTURE: WEB OF RELATIONS

Now, with hierarchical formations

$$WR(s, s', n, m, I_n), \quad WP_{\delta\varepsilon}(s, s', n, m, I_n)$$

at hand we define a structure that constitutes the main subject of the book.

Firstly, a structure, denoted WR and called a web of integer relations, that incorporates all the hierarchical formations of integer relations into one whole is defined.

All hierarchical formation of integer relations associated with sequences

$$s = s_1 ... s_n, \; s' = s'_1 ... s'_n \in I_n$$

and their initial segments

$$s(i) = s_1 ... s_i, \; s'(i) = s'_1 ... s'_i \in I_i$$

of length $i = 1, ..., n-1$ are incorporated into one hierarchical formation

$$WR(s, s', m, I_n) = \bigcup_{i=1}^{n} WR(s(i), s'(i), i, m, I_i)$$

of

$$C(s, s') = \max_{i=1,...,n} C(s(i), s'(i), i)$$

levels. Thus the maximum level of a hierarchical formation

$$WR(s, s', m, I_n)$$

equals the structural complexity $C(s, s')$ of a sequence s with respect to a sequence s'.

The structure, also called a web of relations, is defined to incorporate all these hierarchical formations of integer relations into one whole.

Definition 3.3 (Web of Integer Relations). *Let*

$$WR(I_n) = \bigcup_{m \in \mathbf{Z}} \bigcup_{s \in I_n} \bigcup_{s' \in I_n} WR(s, s', m, I_n),$$

$$WR(I) = \lim_{n \to \infty} WR(I_n),$$

$$WR = WR(\mathbf{Z}),$$

where all possible pairs of sequences s, s' are considered and integer alphabet I is the set of all integers $\mathbf{Z}$.

Secondly, by using the isomorphism $\psi_{\delta\varepsilon}$, a structure, denoted $WP_{\delta\varepsilon}$ and called a web of integer patterns, that for given δ and ε incorporates all the hierarchical formations of integer patterns into one whole is defined. Namely, the structure in geometrical terms is given by the following definition.

Definition 3.4 (Web of Integer Patterns). *Let $WP_{\delta\varepsilon}$ be defined by the isomorphism*

$$\psi_{\delta\varepsilon} : WR \to WP_{\delta\varepsilon}.$$

Due to the isomorphism, the two structures WR and $WP_{\delta\varepsilon}$ can be viewed as one structure, called a web of relations and denoted **WR**. Elements of the web of relations **WR** have dual character and can be equally seen as integer relations or integer patterns (see Figure 23 as an illustration).

The web of relations **WR** is defined as a new mathematical structure whose elements on one side are relations between integers organized in a certain manner. It is quite evident that the web of relations was bound to manifest itself in one way or another in many cases before. This, of course, takes place and we can easily point to its manifestations that reveal elements of the web of relations in a hidden form.

A suitable example to demonstrate this event is connected with the well-known Pythagorean triple (3,4,5). The celebrated relation between integers 3, 4, 5

$$3^2 + 4^2 = 5^2 \tag{3.54}$$

can be viewed as a broken ground that results from a "symmetry breaking" of an element of the web of relations. Namely, consider

$$WR(\eta(8), \bar{\eta}(8), -3, B_8), \quad WP_{11}(\eta(8), \bar{\eta}(8), -3, B_8),$$

where $m = -3, \delta = 1, \varepsilon = 1$. We have corresponding hierarchical formations of integer relations between $-2, -1, 0, +1, +2, +3, +4, +5$ and integer patterns as shown in Figure 29. The system (3.1) in this case, when 2 is cancelled out, is

$$+5^0 - 4^0 - 3^0 + 2^0 - 1^0 + 0^0 + (-1)^0 - (-2)^0 = 0$$

$$+5^1 - 4^1 - 3^1 + 2^1 - 1^1 + 0^1 + (-1)^1 - (-2)^1 = 0$$

$$+5^2 - 4^2 - 3^2 + 2^2 - 1^2 + 0^2 + (-1)^2 - (-2)^2 = 0. \qquad (3.55)$$

In Figure 29 the nature of these integer relations is highly ordered and symmetric. Their underlying laws are simple and unified. Then through a sequence of "symmetry-breaking", we can end up with a different situation. In particular, rewrite the last relation of (3.55) as

$$(-2)^2 + 1^2 + 3^2 + 4^2 = (-1)^2 + 0^2 + 2^2 + 5^2 \qquad (3.56)$$

to look it more close to (3.54) and consider a "symmetry breaking" of (3.56) that is initiated by 0 and pairs $(-2, +2), (+1, -1)$ "creation".

Let us observe, while (3.56) algebraically still holds integer 0 and integers $-2, -1, +1, +2$ in pairs $(-2, +2), (+1, -1)$ "escape from it and emerge beyond"

$$\begin{array}{ccccccc} -2 & 1 & & -1 & 0 & 2 & \\ \Uparrow & \Uparrow & & \Uparrow & \Uparrow & \Uparrow & \\ . & . & 3^2 + 4^2 = & . & . & . & 5^2. \end{array}$$

The large symmetry that we have in Figure 29 is hidden now. The web of relations **WR** in this example gives a context that puts (3.54) in a proper place within its corresponding element.

4. WHY THE STRUCTURE IS A FINAL THEORY

We have defined a mathematical structure, called a web of relations, and shown two of its forms: namely, algebraic and geometrical. In the algebraic form the web of relations is based on integers as the single concept and is a set of integer relations organized in a special way. Through this form comes a very important property of the structure as it can be seen as a *final theory* with regard to the existence of the elements and their hierarchical formations. This argument needs a clarification.

A final theory is a theory that has irreducible basic concepts [24]. The primal advantage of a final theory comes from the fact that it can describe natural systems without any need for explanations of the systems in terms of deeper concepts.

Ultimate building blocks of the web of relations are integers. They initiate hierarchical formations, level by level, producing more complex elements, which are nothing but integer relations. The web of relations is a unified collection of these hierarchical formations.

In particular, the existence of an element of the web of relations belonging to a level is completely determined by elements belonging to the previous level. These elements are true integer relations and when combined by the organizing principle produce a true integer relation. By a true integer relation is meant that the constituent expression of the relation equals zero. This expression, which may be seen as a code of the relation, involves integers subject to arithmetic rules, i.e., addition, subtraction and multiplication, and the existence of the element-relation is completely specified by the fact that upon using the rules this expression can be reduced to zero. If the expression equals zero, then the elements of the previous level form the element; if the expression does not equal zero, then the corresponding elements can not form an element on the next level.

Results in section 2 of this chapter show that integer relations admit "incarnation" as geometrical objects and consequently are more than a set of abstract symbols. But importantly the existence of the integer relations are completely determined by rules of arithmetic, i.e., by the integers.

This property of elements of the web of relations may be helpful in answering an important question: why some elements in nature give rise to a specific element. In particular, suppose the web of relations describes a natural phenomenon in which elements are formed in accordance with hierarchical formations of elements of the web of relations. Then there is no need to look for a deeper principle to explain the formation of elements in the natural phenomenon. The formation is completely specified by the integers and does not need a deeper principle for its explanation unless we change the integers.

We consider an example to illustrate that integer relations as elements of the web of relations have a different meaning than we are used to. Usually the arithmetic expression

$$+5^2 - 4^2 - 3^2 + 2^2 - 1^2 + 0^2 + (-1)^2 - (-2)^2 \tag{3.57}$$

is seen as a record of operations to be applied to the integers with the result of these operations being of the main interest. By using rules of arithmetic it turns out that

$$+5^2 - 4^2 - 3^2 + 2^2 - 1^2 + 0^2 + (-1)^2 - (-2)^2 = 0, \qquad (3.58)$$

and then this can be effectively used to replace (3.57) by 0 in further writings.

It can be shown that (3.58) is an element of the structure. However, the reading of (3.58) in our case is different from the previous one. The left-hand side of (3.58) even does not look conventional as it contains zero, which is usually skipped in computational writings.

In our case it is important because in essence the integer relation (3.58) stands for an element of the structure. Put it in another way (3.58) is an algebraic form of an element of the structure. This form explains why the element exists and gives the code of how it is formed in the structure. The explanation comes by the fact that the left-hand side of (3.58), i.e., the constituent expression, equals zero.

Therefore, as soon as the integers are accepted as the single assumption the web of relations appears automatically, because the existence of its elements and their hierarchical formations are completely specified by the integers.

The integers are considered as the most fundamental entities irreducible to something more simpler. Consequently, the web of relations can be viewed as a final theory with regard to the existence of its elements and their hierarchical formations. This motivates us to propose the web of relations as an universal scale to measure the complexity of natural systems in terms of the hierarchical formations of integer relations.

The geometrical form of the web of relations develops a new perception of integer relations as two-dimensional geometrical objects that can form into each other. This makes it possible to assign physical quantities to elements of the web of relations, which may be used as parameters in description of natural systems. For example, the area of an integer pattern may be seen as an important physical quantity of the element.

The geometrical form also reveals a picture of hierarchical formations and our intuition leads us to believe that they have nonlocal order and large symmetry, as we observed in the example of section 6 in Chapter 2 and many other examples not included in the text. These hierarchical formations point to a clear analogy in nature, where many natural phenomena can be interpreted as hierarchical formations of more elementary phenomena that occur at different scales.

With the web of relations at hand, clarification is then needed as to how it is used in understanding universal principles of emergent computation. The next chapter describes the situation.

Chapter 4

THE STRUCTURE AND EMERGENT COMPUTATION

In Chapter 3 it was shown that an attempt to capture the unified whole in mathematical terms resulted in the discovery of a structure, which may be seen as a final theory. The structure does not rely on any information about space-time. Surprisingly or not it is based on integers as the single assumption. The structure is also called a web of relations because it may be viewed as a web of routes through which hierarchical formations of its elements evolve.

Hierarchical formations of the web of relations, level by level, produce more complex elements and their character is very close to what is associated with complexity. This motivates us to propose the structure as an universal scale to measure the complexity of natural systems in terms of the hierarchical formations of integer relations. Based on these hierarchical formations a concept of structural complexity is defined.

In this Chapter the web of relations as a final theory is used to probe universal principles of emergent computation. In section 1 a principle is presented to discuss a variety of issues connected with the mathematical specification of natural systems in terms of hierarchical formations of the web of relations. The principle cannot be derived from the web of relations only and other means are needed for its justification. However, it is possible to discern outlines of the principle as the hierarchical formations have the preferred bottom-up direction of growth. This suggests that hierarchical formations achieving the maximum level among possible ones stand out as special. This observation accords well with the recognition that the unfolding of ever-greater complexity may be a fundamental property of nature. This recognition is realized in defining the above principle in a concrete mathematical form by using the concept of structural complexity.

In section 2 to study the principle, a model that captures emergent computation and provides abstractions for the development of optimization algorithms is considered. In the model the thinking of decision makers is an integral part of the situation in which they participate. This limits the use of the standard approach in which natural systems are viewed as independent and separable.

In section 3 by using the model the principle is expressed in terms of a notion of coherent solution. The notion says that binary sequences of a coherent solution

maximize structural complexity. The coherent solutions in Chapter 5 are used to present new results in emergent computation.

—

1. HIERARCHICAL FORMATIONS AND PROBING UNIVERSAL PRINCIPLE OF EMERGENT COMPUTATION

In Chapter 3 it was shown that an attempt to capture the unified whole in mathematical terms resulted in the discovery of a structure, which may be seen as a final theory. In essence, this attempt encapsulates the idea that the world is a reflection of this undivided, interconnected structure of integer relations and evolves according to its laws. The structure does not rely on any information about space-time. Surprisingly or not it is based on integers as the single assumption. The structure is also called a web of relations because it may be viewed as a "web" of routes through which hierarchical formations of its elements evolve.

The structure has two forms, i.e., algebraic and geometrical. In the algebraic form elements of the structure are integer relations which simply are and involve in their definition no other concepts but integers. The geometrical form complements the algebraic one and incarnates the elements in terms of two-dimensional geometrical objects.

Therefore, the algebraic form makes the structure a final theory while the geometrical one allows us to perceive the elements as real entities, whose geometrical properties can be used to describe natural systems.

In this chapter the web of relations as a final theory is used to probe universal principles of emergent computation. In this section a principle is presented to discuss a variety of issues connected with the mathematical specification of natural systems in terms of hierarchical formations of the web of relations.

For example, in classical mechanics the behaviour of a natural system is specified by Newton's laws of motion. Similarly, we are interested in understanding how the behaviour of a natural system is reflected in the structure in terms of hierarchical formations and how they are specified.

It is becoming generally recognized that the unfolding of ever-greater complexity may be a fundamental property of nature (for example, [30]). We use this fact as a guide to specify natural systems in terms of the web of relations and try to understand the principle from hierarchical formations of the structure as they are. It is possible to discern outlines of the principle as the hierarchical formations have the preferred bottom-up direction of growth.

Firstly, we have shown that a natural system admits a description in terms of the web of relations. There appears little doubt about the existence of hierarchical formations of the structure. Elements of a hierarchical formation as true integer relations do exist, they simply are. As a hierarchical formation evolves it may be seen as if it takes one possible route among many others admitted by the structure.

Imagine that we observe a hierarchical formation by representing its elements as integer patterns. Suppose that only those integer patterns that are already involved in the formation are visible. At this moment we can see only trees whose elements on different levels give a static picture containing the history of this hierarchical formation. But this picture can change.

It becomes dynamic if a corresponding natural system proceeds in a manner leading to the formation of new elements as the compositions of some visible elements and the new elements also become visible on the next levels. As a result we observe a visible wave in the structure. The wave starts with integers on the ground level and elevates through levels.

Of course, the new elements always exist in the structure and are made visible to show which route and direction the hierarchical formation takes in the structure. Importantly, the compositions of elements and appearance of new elements are completely specified in the structure by the laws of integers and no other rules are applied.

Thus, as a natural system develops in space-time, we observe in the web of relations a process that is close to our intuitive understanding about natural hierarchical formation. The process starts on the ground level and new elements on higher levels appear as the compositions of elements from previous levels.

We see that hierarchical formations of the structure have one important feature. They grow across levels with this growth having a specific bottom-up direction. Figuratively speaking hierarchical formations of the structure strive to elevate as high as possible. This suggests that hierarchical formations in the structure achieving the maximum level among possible ones stand out as special. This observation resonates well with the above mentioned recognition that the unfolding of ever-greater complexity may be a fundamental property of nature.

Secondly, we have every reason to characterize the hierarchical formations in terms of complexity. Indeed, a hierarchical formation of the structure is a process that produces elements on a level as compositions of elements from the lower level. The process also can be seen in the reverse order, as an element can be decomposed into elements of the previous level. The decomposition can be applied again to the elements and repeated so until elements of the ground level are produced.

Then it becomes natural to propose the following:

Definition 4.1 (Complexity of Element of the Structure). *The complexity of an element of the structure equals the number of the level to which the element belongs.*

Definition 4.1 captures that an element of a level is successively formed, level by level, from elements belonging to levels below this level.

A hierarchical formation of the structure is a collection of trees of its elements. In a hierarchical formation the top element of a tree is an element that can not be combined with other elements of the hierarchical formation on the same level to form an element on the next one.

The top element of a tree of a hierarchical formation can be viewed as the result of the hierarchical formation with the tree seen as a part of it. This suggests as a first step in characterizing the complexity of a hierarchical formation of the structure to propose the following definition.

Definition 4.2 (Complexity of Hierarchical Formation). *The complexity of a hierarchical formation $WR(s, s', m, I_n)$ of the structure equals the maximum of complexities of its elements.*

From Definition 4.2 it easily follows that the complexity of a hierarchical formation $WR(s, s', m, I_n)$ is its maximum level, which in turn equals the structural complexity $C(s, s')$ of corresponding sequences s, s'.

Thus structural complexity $C(s, s')$ turns out to be a measure of complexity of a hierarchical formation of the web of relations. This explains once again why we used the name "structural complexity" for $C(s, s')$ when the concept was defined in Chapter 2.

Now we present the above principle in a concrete mathematical form by using the concept of structural complexity.

Principle. *A natural system specified in terms of the web of relations comprises hierarchical formations that have maximum possible structural complexity.*

The principle can not be derived from the web of relations only and other means are needed for its justification. However, we stress once again that it is possible to discern outlines of the principle as the hierarchical formations have the preferred bottom-up direction of growth. This suggests that hierarchical formations achieving the maximum level among possible ones stand out as special and the principle captures this fact.

The principle figuratively says that "nature realizes structure with maximum complexity". Since the days of Darwin, evolution has been associated with in an increase in complexity.

What constitutes the principal difference is that in our case the definition of complexity is formally rigorous and, crucially, is backed up by a final mathematical structure.

2. A MODEL OF COMPUTATION RELEVANT TO THE UNIVERSAL PRINCIPLES

In this section to study the principle a model is presented. The model captures emergent computation and provides abstractions for the development and analysis of optimization algorithms. In the model the thinking of decision makers is an integral part of the situation in which they participate. This limits the use of the standard approach in which natural systems are viewed as independent and separable.

The model can be described as follows. There is a group of decision makers who seek to understand and use in a proper way the situation in which they participate. Their understanding serves as the basis for decisions they have to make in successive steps. These decisions at each step create an event that provides a result of their actions. Expectations of some of them are validated by the result, others are not. Then each decision maker independently employs the event they created together to understand the situation. At each step, a decision maker updates his or her understanding of the situation and incorporates this into the next decision. As a result the global behaviour of the group emerges and is of the main interest in the model.

Interestingly, the thinking of decision makers through their decisions affects the course of events and the course of events affects their thinking. The thinking and the course of events cannot be considered separately and independently. Consequently, this limits the use of the standard approach in which natural systems are viewed as independent and separable. This makes it natural to apply the web of relations to the problem.

The El Farol problem, suggested by W.B. Arthur [31], provides a concrete specification for the model. The problem captures emergent computation and is illuminating to our intuitions.

In particular, the problem can be described as follows. Assume there is a group of $N \geq 2$ people, who enjoy going to El Farol, a small bar, to listen to the music on Thursday evenings. However, none of them wants to visit the bar if it is going to be crowded.

For example, one person may consider the bar to be crowded when more than half of the people are present whilst another one may consider the bar to be crowded when it is full to capacity.

As a result each person seeks to understand and use in a proper way the problem situation in which he or she participates. In particular, on Thursday each person, before going to the bar, employs an algorithm to predict if it will be crowded at the bar in the evening. It is assumed that each person only has the information of how many people were in the bar over all past weeks.

Each person decides independently to go to the bar only if his or her prediction is that it will be not crowded. Because different people have different algorithms and different meanings of whether "the bar is crowded", some will turn up at the bar, while others expect it will be crowded and do not go there. While the number of people going to the bar is uncertain, there is no uncertainty in each person's decision, he or she takes an action: to go or not to go to the bar.

These decisions create a result, the emergent number of people at the bar on the Thursday evening as some will turn up at the bar, while others expect it will be crowded and do not go there. As the new attendance number becomes known on the next day, the expectations of some of them are validated by the result, others are not.

For example, someone may have gone to the bar and found it not crowded, while another person may have stayed at home even though the bar was not crowded. Then each person updates the accuracy of his or her algorithm for the next Thursday. Therefore, each person's decision making affects the problem situation, and the problem situation affects the decision making. They can not be considered separately and independently.

Each person is characterized by an algorithm that is supposed to predict if it will be crowded at the bar on Thursday evenings or not. This prediction is used by the person as the basis for the decision to go or not to go to the bar. Therefore, the algorithm takes only values crowded and not crowded. Symbols -1 and $+1$ can be used respectively for denotation of the values.

Namely, if the algorithm predicts that it will be crowded in the bar then the value is -1. This means that the person does not go to the bar. If it predicts that it will not be crowded in the bar then the value is $+1$. This means that the person does not stay at home and goes to the bar.

Accordingly, the result, due to the attendance number, can be encoded by symbols -1 and $+1$. If an attendance number is interpreted by the person that it was crowded at the bar then the result is -1 and $+1$ otherwise.

An algorithm A for a period of n weeks $n \geq 2$ is a set

$$\{\chi_k, k = 1, ..., n\}$$

of rules. Each rule

$$\chi_k, \quad k = 1, ..., n$$

in general, uses all information given by the past $k-1$ weeks and takes $s_k \in \{-1, +1\}$ as the prediction of a result $s'_k \in \{-1, +1\}$ at the kth week.

This information can be encoded as sequences

$$s(k-1) = s_1...s_{k-1}, \quad s'(k-1) = s'_1...s'_{k-1},$$

where

$$s_i, s'_i \in \{-1, +1\}, \quad i = 1, ..., k-1$$

are a prediction and result at the ith step respectively when $k = 2, ..., n$, and

$$s(0) = \emptyset, \quad s'(0) = \emptyset,$$

where $\emptyset$ is the empty sequences when $k = 1$.

Hence, for a rule $\chi_k, k = 1, ..., n$ of an algorithm A we have

$$s_k = \chi_k(s(k-1), s'(k-1)). \tag{4.1}$$

The behaviour of a group of N people during n weeks can be described as a set of N binary sequences of length n

$$S = \{\bar{s}_i = s_{i1}...s_{in}, i = 1, ..., N\},$$

where a sequence

$$\bar{s}_i = s_{i1}...s_{in} \in S$$

specifies the behaviour of the ith $i = 1, ..., N$ person with $s_{ij} = +1$ if the person goes to the bar on the jth $j = 1, ..., n$ week and $s_{ij} = -1$ otherwise.

The model is a version of the El Farol problem specified by the following condition. The group must interact in such a way that the *emergent number* of people each Thursday turning up at the bar must be equal to the number of seats N' in it.

In the version for each person of the group the term "crowded" means that there are no spare seats at the bar. As the number of seats changes the group must adapt accordingly, self-organizing itself to the new condition.

There are several reasons why the El Farol problem was used apart from being very illuminating to our intuition.

Firstly, the group of people in the El Farol problem exhibits collective behaviour that emerges from their individual decision making rules. The problem easily fits in the framework of emergent computation.

In particular, computing the number of seats in the bar is a useful information processing ability of the group. It is easy to appreciate how lucky the owner of the bar would be to have this group of people as customers. Each time for whatever reasons he or she changes the number of the seats, the bar, without any restructuring cost, has exactly as many people as it can accommodate.

Secondly, the behaviour of the group in the El Farol problem can be completely characterized by a matrix consisting of binary sequences. This gives a clear representation in terms of hierarchical formations of the web of relations and links with different areas expected to be helpful in future.

Thirdly, the problem has a social aspect which can be tested by the principle. Hidden behind binary sequences of the problem's description, there is a social process involving people and their interactions. This process may be interpreted by the people in categories that involve more moral values rather than formal ones. Participating in the process people may be concerned about equal opportunities, responsibility and other moral issues. But these categories may be also applied to interpret solutions given by the principle. For example, it would be interesting to know if everyone visits the bar regularly or do they visit according to a preference.

3. THE PRINCIPLE IN THE EL FAROL PROBLEM AND COHERENT SOLUTIONS

Let $P(n, N, N')$ be a set of solutions to the version of the El Farol problem when for n weeks the group of N people visit the bar with N'

seats. If a collection of binary sequences

$$S = \{\bar{s}_i = s_{i1}...s_{in} \in B_n, i = 1, ..., N\}.$$

is a solution to the problem $S \in P(n, N, N')$ then for each jth week $j = 1, ..., n$

$$\sum_{i=1}^{N}(1/2)(s_{ij} + 1) = N',$$

where a sequence

$$\bar{s}_i = s_{i1}...s_{in} \in B_n$$

describes the visitation of the ith person $i = 1, ..., N$ of the group. The solution S can be represented as a $n \times N$ matrix

$$M(S) = \{s_{ij}\}, \quad i = 1, ..., N, \quad j = 1, ..., n.$$

A solution $S \in P(n, N, N')$ in the web of relations can be associated with a collection of hierarchical formations

$$WR(S)$$

$$= \{WR(\bar{s}_i, \bar{s}_j, 0, B_n), \ \bar{s}_i, \bar{s}_j \in S, \ i, j = 1, ..., N, i \neq j\} \subset \mathbf{WR},$$

where a hierarchical formation

$$WR(\bar{s}_i, \bar{s}_j, 0, B_n)$$

of the collection $WR(S)$ is connected with the ith and jth person of the group.

The collection $WR(S)$ of hierarchical formations can be considered as the description of *emergent computation* resulting from a collective behaviour of the group.

Therefore, there are two related descriptions of the group in space-time and in the web of relations

$$\left\{ \begin{array}{c} \textit{Space} - \textit{time} \\ \textit{Matrix } M(S) \end{array} \right\}$$

$$\Longrightarrow \left\{ \begin{array}{c} \textit{Web of Relations} \\ \textit{Collection of Hierarchical Formations } WR(S) \end{array} \right\}$$

For a solution $S \in P(n, N, N')$ by $\Omega(S, k)$ denote the number of all pairs $(\bar{s}, \bar{s}')$ with sequences $\bar{s}, \bar{s}' \in S$ corresponding to two different people of the group and having at least structural complexity k. Let

$$\min(S) = \min_{\bar{s}, \bar{s}' \in S} C(\bar{s}, \bar{s}'),$$

$$\max(S) = \max_{\bar{s},\bar{s}' \in S} C(\bar{s},\bar{s}').$$

The model translates the principle into a notion of coherent solution.

Definition 4.3. (Coherent Solution). *A solution $S_0 \in P(n,N,N')$ is called a coherent solution if for any solution $S \in P(n,N,N')$ other than a coherent one*

1. *if* $\min(S_0) = \max(S_0)$ *then*

$$\Omega(S,i) < \Omega(S_0,i), \quad i = \min(S_0)$$

2. *if* $\min(S_0) < \max(S_0)$ *then there exists an integer*

$$0 \le l \le \max(S_0) - \min(S_0) - 1$$

such that

$$\Omega(S,i) = \Omega(S_0,i), \quad i = \min(S_0), ..., \min(S_0) + l$$

and

$$\Omega(S,i) < \Omega(S_0,i), \quad i = \min(S_0) + l + 1, ..., \max(S_0).$$

We illustrate Definition 4.3 by simple examples. Consider the version of the El Farol problem when the bar is open for $n = 5$ weeks, the group consists of $N = 5$ people and the number of seats at the bar is $N' = 5, 4, 3$. Coherent solutions are treated up to the group of permutations of the people.

When $N' = 5$ there are enough seats for all people of the group. A person of the group can visit the bar each week and enjoy the music without taking care about visitations of the other people. There is only one solution S to the problem, which is also the coherent one $S = S_0$

$$\bar{s}_1 = +1+1+1+1+1, \quad \bar{s}_2 = +1+1+1+1+1,$$

$$\bar{s}_3 = +1+1+1+1+1, \quad \bar{s}_4 = +1+1+1+1+1,$$

$$\bar{s}_5 = +1+1+1+1+1,$$

or

$$M(S_0) = \begin{pmatrix} +1 & +1 & +1 & +1 & +1 \\ +1 & +1 & +1 & +1 & +1 \\ +1 & +1 & +1 & +1 & +1 \\ +1 & +1 & +1 & +1 & +1 \\ +1 & +1 & +1 & +1 & +1 \end{pmatrix} \qquad (4.2)$$

Looking at the matrix (4.2) it may seem that this coherent solution has a perfect order. It may be said that the people visit the bar in lines or that the "+1"s of the matrix, interpreted as spins, look in one direction.

However, this order can be viewed as the simplest one because the structural complexity of the coherent solution S_0 is 0. There are no hierarchical formations in the web of relations corresponding to the coherent solution S_0.

When $N' = 4$ it can be shown that there is only one coherent solution S_0. The solution is

$$\bar{s}_1 = +1+1+1+1-1, \quad \bar{s}_2 = +1+1+1-1+1,$$

$$\bar{s}_3 = +1+1-1+1+1, \quad \bar{s}_4 = +1-1+1+1+1,$$

$$\bar{s}_5 = -1+1+1+1+1.$$

or

$$M(S_0) = \begin{pmatrix} +1 & +1 & +1 & +1 & -1 \\ +1 & +1 & +1 & -1 & +1 \\ +1 & +1 & -1 & +1 & +1 \\ +1 & -1 & +1 & +1 & +1 \\ -1 & +1 & +1 & +1 & +1 \end{pmatrix}$$

Sequences of the coherent solution S_0 have the same first structural numbers. Among all possible solutions $P(5,5,4)$ to the problem the coherent solution is the unique one in which all sequences have the same first structural numbers.

In particular, this gives $\min(S_0) = 2$ and it can be seen that $\max(S_0) = 2$. No segments of sequences of a solution $S \in P(5,5,4)$ have the same first structural numbers all together. This gives $\Omega(S,2) < \Omega(S_0,2)$.

Sequences of the coherent solution $S_0 \in P(5,5,4)$ produce in the web of relations hierarchical formations that all have structural complexity 2. According to the coherent solution S_0 each person visits the bar four times and meets any person of the group at the bar the same number of times. We may say that under the coherent solution people in the group have many equal opportunities.

When $N' = 3$ it can be shown that there are three coherent solutions. It is worthwhile to present all of them. With the order of no importance the first coherent solution is

$$\bar{s}_1 = +1+1-1-1+1, \quad \bar{s}_2 = +1-1+1+1-1,$$

$$\bar{s}_3 = +1-1-1+1+1, \quad \bar{s}_4 = -1+1+1-1+1,$$

$$\bar{s}_5 = -1+1+1+1-1,$$

or

$$M(S_0) = \begin{pmatrix} +1 & +1 & -1 & -1 & +1 \\ +1 & -1 & +1 & +1 & -1 \\ +1 & -1 & -1 & +1 & +1 \\ -1 & +1 & +1 & -1 & +1 \\ -1 & +1 & +1 & +1 & -1 \end{pmatrix}$$

The second coherent solution is

$$\bar{s}_1 = +1 - 1 + 1 - 1 + 1, \quad \bar{s}_2 = -1 + 1 + 1 + 1 - 1,$$

$$\bar{s}_3 = +1 - 1 - 1 + 1 + 1, \quad \bar{s}_4 = -1 + 1 + 1 - 1 + 1,$$

$$\bar{s}_5 = +1 + 1 - 1 + 1 - 1,$$

or

$$M(S_0) = \begin{pmatrix} +1 & -1 & +1 & -1 & +1 \\ -1 & +1 & +1 & +1 & -1 \\ +1 & -1 & -1 & +1 & +1 \\ -1 & +1 & +1 & -1 & +1 \\ +1 & +1 & -1 & +1 & -1 \end{pmatrix}$$

The third coherent solution is

$$\bar{s}_1 = +1 - 1 + 1 - 1 + 1, \quad \bar{s}_2 = -1 + 1 + 1 + 1 - 1,$$

$$\bar{s}_3 = +1 + 1 - 1 - 1 + 1, \quad \bar{s}_4 = +1 - 1 + 1 + 1 - 1,$$

$$\bar{s}_5 = -1 + 1 - 1 + 1 + 1,$$

or

$$M(S_0) = \begin{pmatrix} +1 & -1 & +1 & -1 & +1 \\ -1 & +1 & +1 & +1 & -1 \\ +1 & +1 & -1 & -1 & +1 \\ +1 & -1 & +1 & +1 & -1 \\ -1 & +1 & -1 & +1 & +1 \end{pmatrix}$$

A coherent solution S_0 may be seen to consist of three subgroups of people

$$(\bar{s}_1, \bar{s}_2), \quad (\bar{s}_3, \bar{s}_4), \quad \bar{s}_5.$$

All sequences of a coherent solution have the same first structural numbers. In particular, this means that each person visits the bar the same number of times. However, there are other solutions $S \in P(5,5,3)$ to the problem that have this property, i.e.,

$$\Omega(S, 2) = \Omega(S_0, 2).$$

What distinguishes a coherent solution S_0 is that sequences in subgroups $(\bar{s}_1, \bar{s}_2), (\bar{s}_3, \bar{s}_4)$ have the same second structural numbers

$$\vartheta_2(\bar{s}_1) = \vartheta_2(\bar{s}_2), \quad \vartheta_2(\bar{s}_3) = \vartheta_2(\bar{s}_4).$$

It can be shown that

$$\min(S_0) = 2, \quad \max(S_0) = 3$$

and for a solution $S \in P(5,5,3)$ not a coherent one

$$\Omega(S,3) < \Omega(S_0,3),$$

which is what Definition 4.3 claims.

All pairs of sequences of a coherent solution $S_0 \in P(5,5,3)$ (except $(\bar{s}_1, \bar{s}_2), (\bar{s}_3, \bar{s}_4)$) produce in the web of relations hierarchical formations of structural complexity 2. Besides pairs $(\bar{s}_1, \bar{s}_2), (\bar{s}_3, \bar{s}_4)$ give hierarchical formations of structural complexity 3.

Comparing the structural complexities of hierarchical formations of the coherent solutions for $N' = 4$ and $N' = 3$ it can be said that the coherent solution in the first case is less complex than a coherent solution in the second one.

According to a coherent solution $S_0 \in P(5,5,3)$ each person in the group visits the bar the same number of times but he or she meets different people of the group a different number of times.

For example, according to the coherent solution presented the first person meets the second one only once at the bar, while he or she meets the third person twice. It can be shown by computations that in a coherent solution these differences are minimum in comparison with other solutions to the problem. Figuratively, under a coherent solution all people of the group have equal opportunities as far as possible.

Chapter 5

SEARCHING FOR UNIVERSAL PRINCIPLES OF EMERGENT COMPUTATION

In Chapter 4 a principle to probe universal principles of emergent computation was presented. The principle was further expressed in terms of a notion of coherent solution. The notion says that binary sequences of a coherent solution maximize structural complexity or share as many common structural numbers as possible.

In this chapter a first step to find verifications of the principle is made. One of the important discoveries apparent from the study of natural complex systems is that self-organization is closely associated with chaos, which many of them approach through period doubling. The main result of this chapter is that the principle translated in the conventional context of space-time is related to period doubling, an observable phenomenon of self-ogranization in nature. This result can be seen as an experimental verification of the principle.

In particular, to put the translation of the principle in a proper context some known results concerning the symbolic description of period doubling are recalled in section 1.

As the principle seems not subject to the constraints of the Turing model of computation, the problem of how coherent solutions can best be algorithmically approximated is presented in section 2. The approximation problem turns out to be very difficult. It is considered as a special case by reducing it to the problem of binary sequence prediction with structural complexity as the criterion.

An optimal algorithm to the last problem is sought in section 3. The idea is to find a description of the optimal algorithm that can be associated with self-organization. As an approximation two descriptions of the optimal algorithm are given. One description associates the optimal algorithm with the symbolic description of period doubling and shows that period doubling is a manifestation of the principle. This result throws light on a full theory connected with the principle as a candidate for universal principles of emergent computation.

1. PTM SEQUENCE IN THE SYMBOLIC DESCRIPTION OF PERIOD DOUBLING

In this section, to put the principle in a proper space-time context, we recall some known facts concerning the symbolic description of period doubling [43].

The iterations of unimodal functions that map an interval into itself are the simplest models of dynamical systems. These models exaggerate certain properties of dynamical systems to make them more clear and can be described as follows.

Let a unimodal function g be a map that sends the interval $[0,1]$ into itself. There is a real number $\theta \in (0,1)$ such that the function g is monotonically increasing on $[0,\theta)$ and is monotonically decreasing on $(\theta,1]$. Let $g(\theta)=1,\ g(1)=0$.

Iterations of the form

$$x_k \longrightarrow x_{k+1} = g(x_k), \quad k=1,2,... \tag{5.1}$$

are used as a discrete time model of a continuous dynamical system with an index k playing the role of time variable. The symbolic description of (5.1) can be given by a function

$$\gamma(x) = \begin{cases} -1, & x \in [0,\theta] \\ +1, & t \in [\theta,1] \end{cases}$$

which encodes the dynamics of the iterations into a binary sequence.

The behaviour of the symbolic trajectory

$$\gamma(g(x)), \gamma(g(g(x))), ... = \gamma(g^1(x)), \gamma(g^2(x)), ...$$
$$= \gamma(g^k(x)), \quad k=1,2,...$$

is of interest, where $x \in [0,1]$ and $g^k, k=1,2,...$ denote the kth iteration of g.

Consider the lexicographical order on B_n with $-1 < +1$. Let

$$s = s_1...s_n, \quad s' = s'_1...s'_n \in B_n, \quad s \neq s'$$

then $s < s'$ if there exists an integer $i = 1,...,n$ such that

$$s_j = s'_j,\ i=1,...,i-1, \quad s_i < s'_i.$$

Define two operators

$$\mathrm{M} : B_n \longrightarrow B_n,$$

$$\mathbf{S} : B_n \longrightarrow B_n, \quad n = \infty.$$

These operators are the *mod*2 operator and the shift operator. Namely, for

$$s = s_1 ... s_n, \; s' = s'_1 ... s'_n \in B_n$$

let $s' = \mathbf{M}(s)$ if

$$s'_i = (\sum_{j=1}^{i} s_j) mod2, \quad i = 1, ..., n$$

and $s' = \mathbf{S}(s)$ if $s'_k = s_{k+1}, \; k = 1, 2, ...$ for $n = \infty$.

The following result of Allouche and Cosnard [44] connects a description of a dynamical system in terms of an unimodal function with properties of a corresponding symbolic description. These properties are given by using the *mod*2, shift and negation operators.

Let $s = s_1 s_2 ... \in B_\infty$ and $s_1 = +1, s_2 = -1$ then there exists an unimodal function g such that

$$s = \gamma(g^k(1)), \; k = 1, 2, ...$$

if and only if for each $k = 1, 2, ...$

$$\bar{\mathbf{M}}(s) \leq \mathbf{S}^k(\mathbf{M}(s)) \leq \mathbf{M}(s), \tag{5.2}$$

where $\mathbf{S}^k$ denote the kth iteration of the shift operator $\mathbf{S}$.

In view of the last result a linearly ordered set of binary sequences is considered

$$U = \{s \in B_\infty, \; \bar{s} \leq \mathbf{S}^k(s) \leq s, k = 1, 2, ... \} \subset B_\infty,$$

whose first three elements $u^1, u^2, u^3 \in U$ with $u^1 < u^2 < u^3$ are

$$u^1 = (+1 - 1)^*, \quad u^2 = (+1 + 1 - 1 - 1)^*,$$

$$u^3 = (+1 + 1 - 1 + 1 - 1 - 1 + 1 - 1)^*,$$

where $(s)^*$ denote that a sequence s repeats itself infinite times.

Consider a sequence

$$u = +1 + 1 - 1 + 1 - 1 - 1 + 1 + 1 - 1 - 1 + 1 - 1 + 1 + 1 - 1 + 1 ... \in B_\infty,$$

which is connected with the PTM sequence $\bar{\eta} \in B_\infty$ starting with -1 by $\bar{\eta} = -1u$.

The following results are valid [44]:

1. $u \in U$.

2. The set

$$\mathbf{F} = U \bigcap [(-1)^*, u[= \{u^k, k = 1, 2, ...\}$$

is linearly ordered, where a sequence u_k is 2^k periodic and

$$\lim_{k \to \infty} u^k = u.$$

3. Let g_μ be a family of unimodular functions admitting a complete Feigenbaum's period doubling cascade for $\mu_* \in [0, \mu]$ then

$$\mathbf{M}(\gamma(g^k_{\mu_*}(1)) = u,$$

when $k = 1, 2, ...$

By putting -1 at the beginning of each sequence

$$u^k \in \mathbf{F}, \quad k = 1, 2, ... ,$$

the set $\mathbf{F}$ becomes a linearly ordered set

$$\mathbf{F}' = \{-1u^k, \ k = 1, 2, ...\} = \{v^k, \ k = 1, 2, ...\}$$

of sequences that are connected with the PTM sequence $\bar{\eta}$ by the condition

$$v^k = (\bar{\eta}(2^k))^*, \quad k = 1, 2, ...$$

Thus period doubling results in the linearly ordered set of sequences

$$\mathbf{F}' = \{v^k, \ k = 1, 2, ...\}.$$

The order in the set specifies a route to a chaotic behaviour, whose symbolic description is given by the PTM sequence $\bar{\eta}$. The route starts at v^1 and sequentially goes from v^k to $v^{k+1}, k = 1, 2, ...$

This order finds explanation in terms of the principle. From Conjecture 2.1 we have that the structural complexity of the sequence

$$v^k(2^k) = \bar{\eta}(2^k), \quad k = 1, 2, ...$$

attains a maximum value among all sequences $s \in B_{2^k}$ except $\eta(2^k)$, i.e.,

$$C(\bar{\eta}(2^k)) = C(\eta(2^k)) = \max_{s \in B_{2^k}} C(s) = k + 1,$$

$$C(\bar{\eta}(2^k)) > C(s), \quad s \in B_{2^k}, \ s \neq \bar{\eta}(2^k), \eta(2^k),$$

where $k = 1, 2, ...$

Consequently, the transition from $v_k = \bar{\eta}(2^k)$ to $v_{k+1} = \bar{\eta}(2^{k+1})$ for each $k = 1, 2, ...$ is specified by the maximization of structural complexity, which is what the principle is about. *Therefore, period doubling turns out to be a manifestation of the principle.*

2. APPROXIMATION TO COHERENT SOLUTIONS AND THE PROBLEM OF BINARY SEQUENCE PREDICTION

In Chapter 4 the principle was expressed in terms of a notion of coherent solution. The notion says that binary sequences of a coherent solution maximize structural complexity. In this section the problem of how coherent solutions can best be approximated algorithmically is presented.

In general, this approximation problem can be formulated as follows. For each ith person $i = 1, ..., N$ find an algorithm A_i such that the group of people as a whole generates solutions to the version of the El Farol problem that are close to coherent solutions in terms of structural complexity as much as possible. This problem is very difficult and it becomes natural at first to consider the problem when there are only two people in the group $N = 2$.

Solutions to the problem take into account two aspects. Firstly, a solution to the problem must satisfy the constraint that the number of people each week visiting the bar equals the number of seats in it. Secondly, two sequences of a solution maximizes the structural complexity. Simplifying the problem we concentrate on the second aspect of it only.

In this case the problem is about a person, say the first one, who, by using an algorithm

$$A = \{\chi_k, k = 1, ..., n\},$$

seeks to produce a sequence $s \in B_n$ having maximum structural complexity with a sequence $s' \in B_n$ generated by the second one. At each kth week $k = 1, ..., n$ the first person uses a rule χ_k and produces the kth symbol

$$s_k = \chi_k(s(k-1), s'(k-1)) \tag{5.3}$$

of the sequence s before the kth symbol s'_k of the sequence s' becomes known.

It is assumed that a value of s'_k is totally indeterminate before the production of s_k, i.e., symbols -1 and $+1$ appear with the same probability. After n steps an algorithm A for a sequence

$$s' = s'_1 ... s'_n \in B_n$$

produces, due to (5.3), a sequence

$$s = s_1 ... s_n \in B_n,$$

denoted by $s = A(s')$.

The problem just described is the problem of binary sequence prediction with structural complexity as the criterion. Usually the problem of binary sequence prediction uses a criterion that is based on the measure $L(s, s')$ given by (2.35). This measure counts the number of components that sequences $s, s' \in B_n$ have in common and comes naturally in the context of space-time. It is worthwhile considering our problem along with the problem of binary sequence prediction when the measure $L(s, s')$ is used as the criterion.

Let $\mathbf{A}$ be the set of all algorithms of type (5.3). Two algorithms

$$A_1 = \{\chi_{1k}, k = 1, ..., n\}, \; A_2 = \{\chi_{2k}, k = 1, ..., n\} \in \mathbf{A}$$

are different if there exists a sequence $s' = s'_1 ... s'_n \in B_n$ and an index $k = 1, ..., n$ such that their predictions at the kth step are not equal

$$\chi_{1k}(s(k-1), s'(k-1)) \neq \chi_{2k}(s(k-1), s'(k-1)),$$

where for $k = 2, ..., n$

$$s_i = \chi_{1i}(s(i-1), s'(i-1)),$$

$$s_i = \chi_{2i}(s(i-1), s'(i-1)), \quad i = 1, ..., k-1.$$

The measure $L(s, s')$ can be used as an algorithm performance measure. For sequences $s, s' \in B_n$ such that $s = A(s')$ it gives the number of times an algorithm A makes a correct prediction about an incoming symbol of the sequence s'. Namely, an algorithm A correctly predicts a sequence $s' = s'_1 ... s'_n \in B_n$ at the kth step $k = 1, ..., n$ if it produces a symbol s_k such that $s_k = s'_k$ and incorrectly predicts the sequence at the kth step if it generates a symbol s_k such that $s_k = -s'_k$.

For an algorithm performance measure two types of analysis to define optimality are considered - the worst case and the average case.

An algorithm A^* with respect to $L(s, s')$ is called optimal in the worst case, if

$$\min_{s' \in B_n, s = A^*(s')} L(s, s') = \max_{A \in \mathbf{A}} \min_{s' \in B_n, s = A(s')} L(s, s')$$

and optimal in the average case, if

$$\sum_{s' \in B_n, s = A^*(s')} L(s, s') = \max_{A \in \mathbf{A}} \sum_{s' \in B_n, s = A(s')} L(s, s'),$$

where the sum is taken over all $s' \in B_n$. The following results describe the situation concerning the optimal algorithms defined with respect to $L(s, s')$.

Lemma 5.1 *Let*

$$A = \{\chi_k, k = 1, ..., n\} \in \mathbf{A}$$

then

$$\min_{s' \in B_n, s = A(s')} L(s, s') = 0.$$

PROOF. We prove by induction on $k = 1, ..., n$ that there exists a sequence $s' = s'_1 ... s'_n \in B_n$ such that $s = A(s')$ and

$$s_k = -s'_k, \quad k = 1, ..., n.$$

Clearly, this means $L(s, s') = 0$, which is what the lemma claims.

For $k = 1$, by definition $s_1 = \chi_1 = const$. Let $s'_1 = -\chi_1 = -s_1$. Since $s_1 = -s'_1$, the result follows. Assume there exists

$$s'(k) = s'_1 ... s'_k, \quad k = 1, ..., n - 1$$

such that

$$s_i = \chi_i(s(i-1), s'(i-1)) = -s'_i, \quad i = 1, ..., k.$$

Then at the $(k+1)$th step of the algorithm A we have

$$s_{k+1} = \chi_{k+1}(s(k), s'(k)).$$

But s'_{k+1} can be such that $s'_{k+1} = -s_{k+1}$. This completes the proof.

Lemma 5.2 *Let*

$$A = \{\chi_k, k = 1, ..., n\} \in \mathbf{A}$$

then

$$\sum_{s' \in B_n, s = A(s')} L(s, s') = n2^{n-1}.$$

PROOF. We first show that the algorithm A correctly predicts 2^{n-1} values of $s'_k, k = 1, ..., n$ when summing over all sequences $s' \in B_n$. Since, for $k = 1$ there are 2^{n-1} sequences $s' = s'_1 ... s'_n$ such that $s'_1 = \chi_1$, where $\chi_1 \in \{-1, 1\}$, then it follows that 2^{n-1} values of s'_1 can be correctly predicted by the algorithm.

Clearly, at the kth, $k = 2, ..., n - 1$ step there are 2^{k-1} different segments $s'(k-1) = s'_1 ... s'_{k-1}$ of all sequences $s' \in B_n$. For a segment $s'(k-1) = s'_1 ... s'_{k-1}$ we have $s(k-1) = s_1 ... s_{k-1}$, where

$$s_i = \chi_i(s(i-1), s'(i-1)), \quad i = 1, ..., k.$$

Since there are 2^{n-k} sequences $s'' = s''_1...s''_n \in B_n$ such that

$$s''_i = s'_i, \quad i = 1, ..., k-1$$

and $s''_k = s_k$, then 2^{n-k} values of s'_k can be correctly predicted by the algorithm A.

Summing over all 2^{k-1} possible segments, we obtain that 2^{n-1} values of s'_i can be correctly predicted at the kth step by the algorithm. Finally, summing over all n steps of the algorithm A we have that the number of correctly predicted values equals $n2^{n-1}$. This completes the proof.

Consequently, from the lemmas it follows that when there is no information about the sequence at all, the measure $L(s, s')$ cannot help us specify optimal algorithms. However, it turns out that this situation allows us to specify optimal algorithms in terms of structural complexity. This serves to illustrate that space-time and the web of relations have different viewpoints on things.

When the problem of binary sequence prediction is considered in terms of structural complexity, an algorithm A can be described by the total structural complexity

$$\sum_{s' \in B_n, s=A(s')} C(s, s') \tag{5.4}$$

associated with the algorithm. The interest is to find an algorithm A_c that maximizes the total structural complexity

$$\sum_{s' \in B_n, s=A_c(s')} C(s, s') = \max_{A \in \mathbf{A}} \sum_{s' \in B_n, s=A(s')} C(s, s'). \tag{5.5}$$

For the version of the El Farol problem and $N = 2$ an optimal algorithm provides in terms of the structural complexity the closest possible approach to coherent solutions and can be seen as the best approximation to the principle given by the algorithms of **A**.

The idea concerning the optimal algorithm is to find its description that can be associated with something observable in nature. This can serve as an experimental verification of the principle. In this respect special attention is paid in order to associate the optimal algorithm with self-organization.

Two general remarks about the optimal algorithm A_c are quite evident. Firstly, the condition

$$\max_{s \in B_n} C(s, s') = C(s'),$$

indicates that for each $s' \in B_n$ an optimal algorithm A_c must strive to produce a sequence $s = A_c(s')$ such that $C(s,s') = C(s')$. Secondly, the contribution to the sum by a sequence s' can be greater if its structural complexity is greater. This implies that an optimal algorithm A_c must be inclined to produce a sequence $s = A_c(s')$ to satisfy the condition $C(s,s') = C(s')$, depending on the value of $C(s')$.

However, it is not clear how to find the optimal algorithm (5.5) immediately. It is proposed then to understand main features of the optimal algorithm by using an approximation to it.

In this approximation for an algorithm A and sequences $s, s' \in B_n$ such that $s = A(s')$ the idea is to consider not the sequences s, s' themselves but only their symbols associated with incorrect steps of the algorithm A. These symbols are viewed then not as separate symbols but as whole sequences. Formally these sequences, denoted by $s \ominus s', s' \ominus s$ and called mistake sequences, are defined as follows.

If there exists an index $k_1, 1 < k_1 \leq n$ such that

$$s_j = s'_j, \quad 1 \leq j \leq k_1 - 1, \quad s_{k_1} = -s'_{k_1}$$

and $s_1 = -s'_1$ for $k_1 = 1$, then $(s \ominus s')_1 = s_{k_1}$. Else, let $s \ominus s'$ be the empty sequence. Assume the first $1 \leq i < n$ symbols of $s \ominus s'$ are defined, i.e.,

$$(s \ominus s')(i) = (s \ominus s')_1 ... (s \ominus s')_i.$$

If there exists an index k_{i+1} such that

$$k_i + 1 < k_{i+1} \leq n$$

and

$$s_j = s'_j, \quad k_i + 1 \leq j \leq k_{i+1} - 1, \quad s_{k_{i+1}} = -s'_{k_{i+1}}$$

with $s_{k_{i+1}} = -s'_{k_{i+1}}$ for $k_{i+1} = k_i + 1$, then $(s \ominus s')_{i+1} = s_{k_{i+1}}$.

Else, let

$$s \ominus s' = (s \ominus s')(i).$$

Set $s' \ominus s$ as the negation of $s \ominus s'$, i.e.,

$$s \ominus s' = \overline{(s' \ominus s)}, \quad (s \ominus s')_i = -(s' \ominus s)_i, \quad i = 1, ..., |(s \ominus s')|. \quad (5.6)$$

Clearly, a sequence $s \ominus s'$ is defined consecutively from incorrect steps with the ith symbol of $s \ominus s'$ corresponding to the ith $i = 1, ..., |s \ominus s'|$ incorrect step. For example, let

$$s = +1 + 1 - 1 + 1 + 1 - 1, \quad s' = +1 - 1 + 1 + 1 - 1 + 1$$

then

$$s \ominus s' = +1 - 1 + 1 - 1, \quad s' \ominus s = -1 + 1 - 1 + 1.$$

By using this approximation structural complexity of sequences s, s' is replaced by structural complexity of mistake sequences $s \ominus s', s' \ominus s$ as it is suggested that

$$C(s, s') \leq C(s \ominus s', s' \ominus s).$$

The performance of an algorithm A in terms of structural complexity then can be characterized by

$$\mathcal{P}(A) = \sum_{s' \in B_n, s = A(s')} C(s \ominus s', s' \ominus s), \tag{5.7}$$

instead of (5.5). An algorithm, denoted by A_c^*, which maximizes

$$\sum_{s' \in B_n, s = A_c^*(s')} C(s \ominus s', s' \ominus s) = \max_{A \in \mathbf{A}} \sum_{s' \in B_n, s = A(s')} C(s \ominus s', s' \ominus s), \tag{5.8}$$

is defined and sought as an approximation to the optimal algorithm A_c.

3. CONSTRUCTING OPTIMAL ALGORITHM AS EXPERIMENTAL VERIFICATIONS OF THE PRINCIPLE

When many options are available a theory is needed to help us understand what might happen and why. Verifications of the theory are expected to come from real observations. This situation is the case with the principle as it needs experimental verifications. As the principle is approximated by an algorithm A_c^*, its verifications can be provided by the algorithm being a mathematical description of an observable natural phenomenon.

In this section two descriptions of the algorithm A_c^* are given. The first description, called structural complexity, shows that rules of the algorithm A_c^* are based on structural complexity. It is important that the second description, called chaotic, is connected with period doubling, an observable phenomenon of self-organization in nature. This result gives an experimental verification of the principle.

Structural complexity description of the algorithm

According to this description the kth rule χ_k^*, $k = 1, ..., n$ of the algorithm A_c^* is based on the structural complexity

$$l = C(s(k) \ominus s'(k), s'(k) \ominus s(k), |s(k) \ominus s(k)'|)$$

of a mistake sequence

$$s(k) \ominus s'(k)$$

with respect to a mistake sequence

$$s'(k) \ominus s(k)$$

and the sign of

$$\vartheta_l(s(k) \ominus s'(k)) - \vartheta_l(s'(k) \ominus s(k)),$$

where

$$s(k) = A_c^*(s'(k)).$$

In this sense it may be said that the algorithm learns from mistakes what decisions to make.

In particular, at the first step mistake sequences are empty and the algorithm A_c^* starts by definition

$$s_1 = \chi_1^*(\emptyset) = +1.$$

At the $(k+1)$th step $k = 1, ..., n-1$ two situations are possible:

1. $s(k) \ominus s'(k) = s(k) \ominus s'(k) = \emptyset$, i.e., mistake sequences are empty,

2. $s(k) \ominus s'(k) \neq \emptyset$, $s'(k) \ominus s(k) \neq \emptyset$, i.e., mistake sequences are not empty.

In the first situation the condition $s(k) \ominus s'(k) = \emptyset$ means that no incorrect steps have been made up to the kth step. Under this circumstance the $(k+1)$th rule χ_{k+1}^* continues to set

$$s_{k+1} = \chi_{k+1}^*(s(k), s'(k))$$

$$= \chi_{k+1}^*(s(k) \ominus s'(k), s'(k) \ominus s(k)) = \chi_{k+1}^*(\emptyset) = +1,$$

where in the notation we stress the character of the rule's dependence.

The second situation is a place where structural complexity is used. Since $s(k) \ominus s'(k) \neq \emptyset$ then

$$s(k) \ominus s'(k) \neq s'(k) \ominus s(k).$$

This means that there exists an integer

$$1 \leq l = C(s(k) \ominus s'(k), s'(k) \ominus s(k), |s(k) \ominus s(k)'|)$$

such that

$$\vartheta_i(s(k) \ominus s'(k)) = \vartheta_i(s'(k) \ominus s(k)), \quad i = 0, ..., l-1$$

but

$$\vartheta_l(s(k) \ominus s'(k)) \neq \vartheta_l(s'(k) \ominus s(k)).$$

Two cases are possible:

1. $\vartheta_l(s(k) \ominus s'(k)) - \vartheta_l(s'(k) \ominus s(k)) > 0,$

2. $\vartheta_l(s(k) \ominus s'(k)) - \vartheta_l(s'(k) \ominus s(k)) < 0.$

The rule χ^*_{k+1} sets the value of s_{k+1} to minimize the difference between the lth structural numbers of mistake sequences

$$s(k+1) \ominus s'(k+1), \quad s'(k+1) \ominus s(k+1).$$

Namely, the rule sets $s_{k+1} = -1$, if

$$\vartheta_l(s(k) \ominus s'(k)) - \vartheta_l(s'(k) \ominus s(k)) > 0,$$

and sets $s_{k+1} = +1$, if

$$\vartheta_l(s(k) \ominus s'(k)) - \vartheta_l(s'(k) \ominus s(k)) < 0.$$

The following notation specifies the character of the rule's dependence

$$\begin{aligned} s_{k+1} &= \chi^*_{k+1}(s(k), s'(k)) \\ &= \chi^*_{k+1}(s(k) \ominus s'(k), s'(k) \ominus s(k)) \\ &= \chi^*_{k+1}(sign(\vartheta_l(s(k) \ominus s'(k)) - \vartheta_l(s'(k) \ominus s(k)))), \end{aligned}$$

where

$$l = C(s(k) \ominus s'(k), s'(k) \ominus s(k), |s(k) \ominus s(k)'|).$$

Clearly, if at the $(k+1)$th step $k = 1, ..., n-2$ it turns out that $s_{k+1} = s'_{k+1}$ then

$$s(k+1) \ominus s'(k+1) = s(k) \ominus s'(k),$$

$$s'(k+1) \ominus s(k+1) = s'(k) \ominus s(k).$$

and the rule χ^*_{k+2} has the same value as the rule χ^*_{k+1}. Thus, if at the kth $k = 1, ..., n-1$ step the rule χ^*_k sets $s_k = \chi^*_k$ then

$$s_{k+i} = \chi^*_{k+i} = s_k = \chi^*_k, \; i = 1, ..., n-k$$

until an incorrect step occurs. In other words if the algorithm A^*_c starts with some value it then continues with the same value until an incorrect step happens.

The algorithm A^*_c has a remarkable property distinguishable by the construction of mistake sequences. The following observation is helpful to explain the property.

Namely, at each kth step $k = 1, ..., n$ a rule χ_k of an algorithm A can not always correctly determine the next symbol s'_k of a sequence s'. Nevertheless, a rule χ_k can always correctly determine the next symbol of a mistake sequences $s(k) \ominus s'(k)$ if the symbol appears at the kth step.

Indeed, if a rule χ_k at the kth step sets $s_k = +1$, but $s'_i = -1$, then the next symbol of a mistake sequence $s(k) \ominus s'(k)$ is $+1$. The mistake sequence $s(k) \ominus s'(k)$ does not change if $s_k = s'_k = +1$. If the rule χ_k sets $s_k = -1$, but $s'_i = +1$, then the next symbol of the mistake sequence $s(k) \ominus s'(k)$ is -1. The mistake sequence $s(k) \ominus s'(k)$ does not change if $s_k = s'_k = -1$.

Therefore, each algorithm A has its own logic for the construction of mistake sequences. The property of the algorithm A^*_c is that it constructs the mistake sequence $s \ominus s'$ in such a way that its kth symbol $k = 1, ..., n$ is the kth symbol of the PTM sequence η. The following theorem demonstrates this property of the algorithm [15].

Theorem 5.1 (Mistakes according to the PTM sequence). *Let*

$$s = s_1 ... s_n, \; s' = s'_1 ... s'_n \in B_n$$

be such that $s = A^*_c(s')$. Then the algorithm A^*_c constructs the mistake sequence $s(k) \ominus s(k)'$ according to the PTM sequence

$$s(k) \ominus s(k)' = \eta(l),$$

$$|s(k) \ominus s(k)'| = l, \quad 1 \leq l \leq k, \quad k = 1, ..., n.$$

PROOF. Properties of the PTM sequence given by Theorem 2.2 is mainly exploited to prove the theorem. Recall

$$\psi_+ = \rho_{011}(\eta(n)) \in W_{11}[t_0, t_n],$$

$$\psi_- = \rho_{011}(\bar{\eta}(n)) \in W_{11}[t_0, t_n].$$

It is evident, the theorem will be proved if we show that there exists a sequence of steps $k_{2^i}, i = 1, 2, ..$ such that, if

$$|s(k_{2^i}) \ominus s'(k_{2^i})| = 2^i,$$

then

$$s(k_{2^i}) \ominus s'(k_{2^i}) = \eta(2^i). \tag{5.9}$$

We prove (5.9) by induction on $i = 1, 2, \ldots$. For the base case $i = 1$ by definition of the algorithm we have

$$s_1 = \chi_1^* = +1.$$

It implies that the first symbol in the mistake sequence $s \ominus s'$ can be only $+1$.

Indeed, all successive rules $\chi_k^*, k > 1$ have this value as long as $s_k = s'_k$, where $s_k = \chi_k^* = +1$. If a k_1th step is the first incorrect one, i.e., $s_{k_1} = -s'_{k_1}$, then we get

$$s(k_1) \ominus s'(k_1) = +1, \quad s'(k_1) \ominus s(k_1) = -1.$$

By definition of structural numbers

$$\vartheta_1(s(k_1) \ominus s'(k_1)) = \psi_+^{[1]}(t_1), \quad \vartheta_1(s'(k_1) \ominus s(k_1)) = \psi_-^{[1]}(t_1),$$

where

$$\psi_+ = \rho_{011}(s(k_1) \ominus s'(k_1)), \quad \psi_- = \rho_{011}(s'(k_1) \ominus s(k_1)).$$

Since

$$\begin{aligned}\vartheta_1(s(k_1) \ominus s'(k_1)) &= \psi_+^{[1]}(t_1) > \psi_-^{[1]}(t_1) \\ &= \vartheta_1(s'(k_1) \ominus s(k_1)),\end{aligned}$$

then according to the rule $\chi_{k_1+1}^*$ its value is -1.

Similar arguments give that the second symbol in the mistake sequence $s' \ominus s$ can be only -1. It happens as soon as at a k_{2^1}th step $s_{k_{2^1}} = -s'_{k_{2^1}}$ and then

$$s(k_{2^1}) \ominus s'(k_{2^1}) = +1 - 1 = \eta(2^1),$$

$$s'(k_{2^1}) \ominus s(k_{2^1}) = +1 - 1 = \bar{\eta}(2^1),$$

which proves the claim for $i = 1$.

Induction hypothesis. Assume (5.9) is true for some integer i, namely

$$s(k_{2^i}) \ominus s'(k_{2^i}) = \eta(2^i), \quad s'(k_{2^i}) \ominus s(k_{2^i}) = \bar{\eta}(2^i),$$

where k_{2^i} is a step at which the 2^ith mistake takes place.

This assumption given we prove that (5.9) is also valid for $i+1$. Consider the interval $[t_0, t_{2^{i+1}})$ and a pair of points

$$t_j, \quad t_{2^i+j}, \quad j = 0, ..., 2^i - 1.$$

We start a second induction within the first one to prove that

$$(s \ominus s')_j = -(s \ominus s')_{2^i+j}, \quad j = 1, ..., 2^i, \tag{5.10}$$

which, since it implies

$$(s \ominus s')(2^{i+1}) = (s \ominus s')(2^i)(s' \ominus s)(2^i),$$

proves the claim for $i+1$.

From Theorem 2.2 we have

$$\psi_+^{[l]}(t_{2^i+j}) = -\psi_+^{[l]}(t_j), \quad \psi_-^{[l]}(t_{2^i+j}) = -\psi_-^{[l]}(t_j), \tag{5.11}$$

where $j = 0, ..., 2^i - 1$, $l = 1, ..., i$. In particular, for the base case $j = 1$ of the second induction (5.11) the theorem gives

$$\vartheta_l(s(k_{2^i}) \ominus s'(k_{2^i})) = \vartheta_l(\eta(2^i)) = \psi_+^{[l]}(t_{2^i}) = 0, \quad l = 1, ..., i,$$

$$\vartheta_l(s'(k_{2^i}) \ominus s(k_{2^i})) = \vartheta_l(\bar{\eta}(2^i)) = \psi_-^{[l]}(t_{2^i}) = 0, \quad l = 1, ..., i$$

and

$$\vartheta_{i+1}(s(k_{2^i}) \ominus s'(k_{2^i})) = \vartheta_{i+1}(\eta(2^i)) = \psi_+^{[i+1]}(t_{2^i}) > 0,$$

$$\vartheta_{i+1}(s'(k_{2^i}) \ominus s(k_{2^i})) = \vartheta_{i+1}(\bar{\eta}(2^i)) = \psi_-^{[i+1]}(t_{2^i}) < 0.$$

This implies that the value of the rule at the $(k_{2^i}+1)$th step is set -1. Hence

$$(s \ominus s')_1 = -(s \ominus s')_{2^i+1},$$

which proves the claim for the base case $j = 1$ of the second induction.

Induction hypothesis of the second induction. Assume that (5.10) is true for $j = 1, ..., 2^i - 1$. This assumption given we prove that it is also valid for $j+1$. Indeed, consider the points t_{2^i+j}, t_j. By the induction hypothesis of the second induction

$$s(k_{2^i+j}) \ominus s'(k_{2^i+j}) = \eta(2^i + j),$$

$$s'(k_{2^i+j}) \ominus s(k_{2^i+j}) = \bar{\eta}(2^i + j),$$

where k_{2^i+j} is a step at which the (2^i+j)th mistake takes place.

Note which structural numbers are used to set the value of the rule at the $(k_{2^i+j}+1)$th step. Since, by Theorem 2.2,

$$\psi_+^{[i]}(t) > 0, \quad \psi_-^{[i]}(t) > 0, \quad t \in (t_{2^i}, t_{2^{i+1}}),$$

then

$$\psi_+^{[i]}(t_{2^i+j}) = \vartheta_i(\eta(2^i+j)) > 0,$$

$$\psi_-^{[i]}(t_{2^i+j}) = \vartheta_i(\bar{\eta}(2^i+j)) < 0.$$

This yields

$$\vartheta_i(\eta(2^i+j)) \neq \vartheta_i(\bar{\eta}(2^i+j)).$$

Therefore, structural numbers

$$\vartheta_{j'}(\eta(2^i+j)), \quad \vartheta_{j'}(\bar{\eta}(2^i+j)), \quad j' = i+1, i+2, \dots,$$

are not considered to set the value of the rule. By definition they are used only when

$$\vartheta_{j'}(\eta(2^i+j)) = \vartheta_{j'}(\bar{\eta}(2^i+j)), \quad j' = 1, \dots, i.$$

Suppose that at the point t_j

$$\vartheta_{l'}(\eta(j)) = \vartheta_{l'}(\bar{\eta}(j)), \quad l' = 0, .., l,$$

$$\vartheta_{l+1}(\bar{\eta}(j)) \neq \vartheta_{l+1}(\eta(j)).$$

Making use of

$$\vartheta_{l'}(\eta(j)) = \psi_+^{[l']}(t_j), \quad \vartheta_{l'}(\bar{\eta}(j)) = \psi_-^{[l']}(t_j), \quad l' = 1, \dots, l+1$$

and (5.11) we come to

$$\vartheta_{l'}(\eta(2^i+j)) = \vartheta_{l'}(\bar{\eta}(2^i+j)), \quad l' = 0, .., l,$$

$$\vartheta_{l+1}(\eta(2^i+j)) \neq \vartheta_{l+1}(\bar{\eta}(2^i+j)).$$

This implies that at the $(k_{2^i+j}+1)$th step the first $l+1$ structural numbers of the mistake sequences are considered. By using (5.11) again, we obtain

$$\begin{aligned}
&\vartheta_{l+1}(\eta(2^i+j)) - \vartheta_{l+1}(\bar{\eta}(2^i+j)) \\
&= \psi_+^{[l+1]}(t_{2^i+j}) - \psi_-^{[l+1]}(t_{2^i+j}) \\
&= -\psi_+^{[l+1]}(t_j) + \psi_-^{[l+1]}(t_j) = -\vartheta_{l+1}(\eta(j)) + \vartheta_{l+1}(\bar{\eta}(j)) \\
&= -(\vartheta_{l+1}(\eta(j)) - \vartheta_{l+1}(\bar{\eta}(j))) \neq 0.
\end{aligned}$$

From the last condition it follows that the value of the rule at the $(k_{2^i+j}+1)$th step is set as the negation of the value of the rule at the (k_j+1)th step.

Consequently, we have

$$(s \ominus s')_j = -(s \ominus s')_{2^i+j}, \quad j = 1, ..., 2^i$$

and

$$(s \ominus s')(2^{i+1}) = \eta(2^{i+1}), \quad (s' \ominus s)(2^{k+1}) = \bar{\eta}(2^{i+1}).$$

This finishes the two inductive proofs simultaneously and completes the proof of the whole theorem.

Chaotic description of the algorithm

Theorem 5.1 provides another description of the algorithm A_c^* and makes it possible to realize it by using a natural process. In particular, suppose there exists a generator $G(\eta)$ of the PTM sequence consequently producing one symbol of the sequence per iteration. The generator can be built on the basis of a dynamical system that has period doubling [44]. With the help of the generator $G(\eta)$ the chaotic description of the algorithm can be represented by the following steps.

Step 1. Set the number of steps $k = 1$ and number of incorrect steps $l = 1$.

Step 2. Generate the lth bit η_l of the PTM sequence η by the generator $G(\eta)$ and for the value of the rule χ_k^* set $\chi_k^* = \eta_l$. Set $s_k = \chi_k^*$.

Step 3. If $s_k = s'_k$, then go to Step 4. Otherwise go to Step 5.

Step 4. If $k = n$, then go to Step 6. Otherwise set $\chi_{k+1}^* = \chi_k^*$ and $s_{k+1} = \chi_{k+1}^*$. Increment the number of steps $k = k+1$ and go to Step 3.

Step 5. If $k = n$, then go to Step 6. Otherwise increment the number of steps $k = k+1$ and the number of incorrect steps $l = l+1$ and go to Step 2.

Step 6. Stop.

The chaotic description of the algorithm A_c^* gives an experimental verification of the principle. For example, when the input sequence s' is the PTM sequence starting with -1 then the algorithm A_c^* is nothing but demonstrates the symbolic description of period doubling, an observable phenomenon of self-organization in nature.

The algorithm A_c^* can be formulated as a simple strategy. This strategy of "win - stay, lose - consult generator $G(\eta)$" considers the generator $G(\eta)$ as a "source of universal wisdom" that can provide the answer to every question. It turns out that the algorithm A_c^* minimizes (5.8) and in this sense the algorithm A_c^* is optimal [17].

Theorem 5.2. *The algorithm A_c^* is the unique algorithm minimizing* (5.8) *among all algorithms of* **A** *starting with* +1.

PROOF. Note that for each algorithm A starting with +1 the number of mistake sequences $s \ominus s'$ such that

$$s' \in B_n, \quad s = A(s'), \quad |s \ominus s'| = i$$

equals $\binom{n}{i}, i = 0, ..., n$.

We prove this statement by induction on $k = 1, ..., n$. For $k = 1$, since any algorithm A starts with $\chi_1 = s_1 = +1$, then there is one mistake sequence $s \ominus s'$ such that $|s \ominus s'| = 1$ when $s_1' = -s_1 = -1$, and one mistake sequence $s \ominus s'$ such that $|s \ominus s'| = 0$ when $s_1' = s_1 = +1$. Since $\binom{1}{0} = 1$ and $\binom{1}{1} = 1$, then the statement is proved for $k = 1$.

Induction hypothesis. Assume the truth of the statement for $k = 1, ..., n-1$. This means that the number of mistake sequences $s \ominus s'$ such that $|s \ominus s'| = i$ equals $\binom{k}{i}, i = 0, ..., k$. This assumption given we prove the statement for $k+1$.

Observe that if at the kth stop a mistake sequence $s(k) \ominus s'(k)$ has length $i = 0, ...k$, i.e., $|s(k) \ominus s'(k)| = i$ then at the $(k+1)$th step this mistake sequence gives rise to two mistake sequences one of which has length i and the other has length $i+1$. Indeed, if case $s'_{k+1} = +1$ occurs, then for $s_{k+1} = +1$ and $s_{k+1} = -1$ we have respectively

$$|s(k+1) \ominus s'(k+1))| = |s(k) \ominus s'(k)| = i$$

and

$$|s(k+1) \ominus s'(k+1)| = |s(k) \ominus s'(k)| + 1 = i + 1.$$

Analogously, if case $s'_{k+1} = -1$ occurs, then we get

$$|s(k+1) \ominus s'(k+1)| = |s(k) \ominus s'(k)| + 1 = i + 1$$

for $s_{k+1} = +1$ and

$$|s(k+1) \ominus s'(k+1)| = |(s(k) \ominus s'(k)| = i$$

for $s_{k+1} = -1$.

In its turn, at the $(k+1)$th step mistake sequences

$$s(k+1) \ominus s'(k+1)$$

such that

$$|s(k+1) \ominus s'(k+1)| = i$$

result from mistake sequences

$$s(k) \ominus s'(k)$$

such that

$$|s(k) \ominus s'(k)| = i, \quad |s(k) \ominus s'(k)| = i - 1.$$

By the induction hypothesis, the number of mistake sequences

$$s(k) \ominus s'(k)$$

such that

$$|s(k) \ominus s'(k)| = i$$

equals

$$\binom{k}{i}, \quad i = 0, ..., k.$$

Therefore, the number of of mistake sequences

$$s(k+1) \ominus s'(k+1)$$

such that

$$|s(k+1) \ominus s'(k+1)| = i$$

equals

$$\binom{k}{i} + \binom{k}{i-1} = \binom{k+1}{i}$$

for $i = 1, ..., k$ and

$$\binom{k}{i}$$

for $i = 0, k+1$. Since

$$\binom{k}{0} = \binom{k+1}{0}, \quad \binom{k}{k} = \binom{k+1}{k+1},$$

then the inductive step is completed and the statement is proved.

Represent the performance of an algorithm A in terms of structural complexity (5.7) as

$$\mathcal{P}(A) = \sum_{i=0}^{n} \sum_{|s \ominus s'| = i, s = A(s')} C(s \ominus s', s' \ominus s), \tag{5.12}$$

where all sequences are grouped together in accordance with their length $i = 0, ..., n$ and then are summed over all possible values.

Let $\mathbf{n} = \{0, 1, ..., n\}$ and $\bar{\mathbf{n}} = \{2, 4, ..., 2^{\lfloor log_2 n \rfloor}\}$. By using Conjecture 2.1, we can rewrite (5.12) in the form

$$\mathcal{P}(A) = \sum_{i \in \bar{\mathbf{n}}} \sum_{|s \ominus s'| = i, s = A(s')} C(s \ominus s', s' \ominus s)$$

$$+ \sum_{i \in \mathbf{n} \setminus \bar{\mathbf{n}}} \sum_{|s \ominus s'| = i, s = A(s')} C(s \ominus s', s' \ominus s)$$

$$< \sum_{i \in \bar{\mathbf{n}}} \sum_{|s \ominus s'| = i, s = A(s')} C(\eta(i), \bar{\eta}(i))$$

$$+ \sum_{i \in \mathbf{n} \setminus \bar{\mathbf{n}}} \sum_{|s \ominus s'| = i, s = A(s')} C(\eta(i), \bar{\eta}(i)). \tag{5.13}$$

It follows from the statement proved above that for any algorithm A we have

$$\sum_{|s \ominus s'| = i, s = A(s')} C(\eta(i), \bar{\eta}(i)) = \binom{n}{i} C(\eta(i), \bar{\eta}(i)). \tag{5.14}$$

By using (5.14), we obtain for (5.13)

$$\mathcal{P}(A) < \sum_{i \in \bar{\mathbf{n}}} \binom{n}{i} C(\eta(i), \bar{\eta}(i)) + \sum_{i \in \mathbf{n} \setminus \bar{\mathbf{n}}} \binom{n}{i} C(\eta(i), \bar{\eta}(i)). \tag{5.15}$$

By Theorem 5.2 and again by the statement proved above, we can rewrite (5.15) as

$$\mathcal{P}(A) < \sum_{i=0}^{n} \binom{n}{i} C(\eta(i), \bar{\eta}(i))$$

$$= \sum_{i=0}^{n} \sum_{|s \ominus s'| = i, s = A_c^*(s')} C(s \ominus s', s' \ominus s) =$$

$$= \sum_{s' \in B_n, s = A_c^*(s')} C(s \ominus s', s' \ominus s) = \mathcal{P}(A_c^*).$$

This completes the proof of the theorem.

References

[1] J. Holland, *Emergence: From Chaos to Order*, Helix Books, 1998.

[2] P. Davies, *The Cosmic Blueprint*, Unwin, London, 1989.

[3] M. Gell-Mann, *The Quark and the Jaguar. Adventures in the Simple and the Complex*, Freeman and Company, New York, 1994.

[4] L. Smolin, *The Life of the Cosmos*, Phoenix, London, 1998.

[5] J. Horgan, *From Complexity to Perplexity*, Scientific American, 1995, June, pp. 74–79.

[6] D. Bohm and B. Hiley, *The Undivided Universe: an Ontological Interpretation of Quantum Mechanics*, Routledge and Kegan Paul, London, 1993.

[7] R. Penrose, *Shadows of the Mind*, Vintage, London, 1995.

[8] D. Bohm, *Wholeness and the Implicate Order*, Routledge and Kegan Paul, London, 1980.

[9] G. Zukav, *The Dancing Wu Li Masters*, Fontana Paperbacks, London, 1982.

[10] N. Bohr, *Atomic Theory and the Description of Nature*, Cambridge, 1961.

[11] W. Heisenberg, *Physics and Philosophy*, Allen and Unwin, London, 1971.

[12] J. Bell, *Speakable and Unspeakable in Quantum Theory*, Cambridge University Press, 1987.

[13] Yu. Evtushenko, V. Korotkich and M. Potapov, "Numerical Methods for Global Optimization", In: C. Floudas and P. Pardalos (eds.), *Recent Advances in Global Optimization*, Princeton University Press, Princeton, 1992, pp. 274–297.

[14] V. Korotkich, *On Structural Properties of Nonlinear Systems in Period Doubling*, Institute for Automated Systems, Soviet Academy of Sciences, Moscow, 1988 (in Russian).

[15] V. Korotkich, *Integer Code Series: Applications in Dynamic Systems and Complexity*, The Computing Center of Russian Academy of Sciences, Moscow, 1993.

[16] V. Korotkich, "A Symbolic Description of Self-Organization in Complex Systems", *Notes of the Russian Academy of Sciences, Journal of Engineering Cybernetics*, Vol. 1, 1994, pp. 20–31.

[17] V. Korotkich, "Multicriteria Analysis in Problem Solving and Structural Complexity", In: P. Pardalos, Y. Siskos and C. Zopounidis (eds.), *Advances in Multicriteria Analysis*, Kluwer Academic Publishers, Dordrecht, 1995, pp. 81–90.

[18] V. Korotkich, "Symmetry in Structural Complexity and a Model of Formation", In: R. Stocker, H. Jelinek, B. Durnota and T. Bossomaier (eds.), *From Local Interactions to Global Phenomena*, IOS Press, Amsterdam, 1996, pp. 84–95.

[19] V. Korotkich, "Symmetry in Structural Complexity and a Model of Formation", *Journal of Complexity International*, Vol. 3, 1996, April.

[20] V. Korotkich, "A Mathematical Framework for Human Decision Making as an Integrated Part of the Whole", In: L. Reznik, V. Dimitrov and J. Kacprzyk (eds.), *Fuzzy Systems Design*, Physica-Verlag, Heidelberg, 1998, pp. 36–62.

[21] V. Korotkich, "On a Structure of Systems of Linear Equations in a Global Description of Sequences", In: K. Holmqvist, S. Migdalas, P. Pardalos and Varband (eds.), *From Local to Global Optimization*, Kluwer Academic Publishers, Dordrecht, 1999.

[22] V. Korotkich, "On a Data Structure in a Global Description of Sequences", In: P. Pardalos, S. Migdalas and Burkard (eds.), *Combinatorial and Global Optimization*, World Scientific Publishing Co, 1999.

[23] V. Korotkich, "On Complexity and Optimization in Emergent Computation", In: P. Pardalos (ed.), *Approximation and Complexity in Numerical Optimization*, Kluwer Academic Publishers, 1999.

[24] S. Weinberg, *Dreams of a Final Theory*, Vintage, London, 1993.

[25] G. Chaitin, *Information, Randomness, and Incompleteness*, World Scientific, 1987.

[26] A. Kolmogorov, *Problems of Information Transmission*, Vol. 1, 1965.

[27] R. Solomonoff, *Information and Control*, Vol. 7, 1964.

[28] C. Bennett, "Logical Depth and Physical Complexity", In: R. Herken (ed.), *The Universal Turing Machine - a Half-Century Survey*, Oxford University Press, 1998, pp. 227–257.

[29] G. Nicolis and I. Prigogine, *Exploring Complexity*, Freeman and Company, New York, 1989.

[30] F. Heylighen and D. Aerts, *The Evolution of Complexity*, Kluwer Academic Publishers, Dordrecht. 1996.

[31] W. Arthur, (1994), "Inductive Reasoning and Bounded Rationality (The El Farol Problem)", *American Economic Review*, Vol. 84, 1994, pp. 406–411.

[32] I. MacDonald, *Symmetric Functional and Hall Polynomials*, The Camelot Press Ltd., Southhampton, 1979.

[33] E. Majorana, "Atomi Orientati in Campo Magnetico Variabile", *Nuovo Cimento*, Vol. 9, 1932, pp. 43–50.

[34] N. Korobov, *Trigonometric Sums and their Applications*, Nauka, Moscow, 1989.

[35] K. Ireland and M. Rosen, *A Classical Introduction to Modern Number Theory*, Springer-Verlag, 1982.

[36] A. Weil, "Numbers of Solutions of Equations over Finite Fields", *Bulletin American Mathematical Society*, Vol. 55, 1949, pp. 497–508.

[37] P. Deligne, "La Conjecture de Weil", *Publ. Math. IHES*, Vol. 43, 1974, pp. 273-307.

[38] R. Hartshorne, *Algebraic Geometry*, Springer-Verlag, 1982.

[39] K. Ireland and M. Rosen, *A Classical Introduction to Modern Number Theory*, Springer-Verlag, New York, 1990.

[40] E. Prouhet, "Memoire sur Quelques Relations entre les Puissances des Nombres", *C.R. Acad. Sci.*, Paris, Vol. 33, 1851, p. 225.

[41] A. Thue, "Uber unendliche Zeichenreihen", *Norske vid. Selsk. Skr. I. Mat. Nat. Kl. Christiana*, Vol. 7, 1906, p. 1.

[42] M. Morse, "Recurrent Geodesics on a Surface of Negative Curvature", *Trans. Amer. Math. Soc.*, Vol. 22, 1921, p. 84.

[43] M. Feigenbaum, "Universal Behaviour in Nonlinear Systems", *Los Alamos Science*, Vol. 1, 1980, pp. 4-27.

[44] J. Allouche and M. Cosnard, "Sequences Generated by Automata and Dynamical Systems", In: *Dynamical Systems and Cellular Automata*, Academic Press, 1985, p. 17.

[45] T. Dantzig, *Number: The Language of Science*, George Allen & Unwin Ltd, London, 1962.

[46] A. van der Poorten, *Notes on Fermat's Last Theorem*, John Wiley & Sons, Inc., New York, 1996.

[47] M. Gardner, *Penrose Tiles to Trapdoor Ciphers*, W. H. Freeman and Company, New York, 1989.

[48] M. Schroeder, *Number Theory in Science and Communication*, Springer Series in Information Sciences, Vol. 7, 1990.

[49] B. Mandelbrot, *The Fractal Geometry of Nature*, Freeman, New York, 1982.

[50] S. Ramanujan, "Note on a Set of Simultaneous Equations", *J. Indian Math. Soc.*, Vol. 4, 1912, pp. 94–96.

[51] L. Mordell, "On a Sum Analogous to a Gauss's Sum", *Quart. J. Math.*, Vol 3, 1932, pp. 161–169.

[52] H. Weyl, " Uber die Gleichverteilung von Zahlen mod", *Eins. Math. Ann.*, Vol. 77, 1916, pp. 313-352.

[53] I. Vinogradov, *Trigonometric Sums Method in Number Theory*, Nauka, Moscow, 1980.

SPRINGER NATURE

GPSR Compliance

The European Union's (EU) General Product Safety Regulation (GPSR) is a set of rules that requires consumer products to be safe and our obligations to ensure this.

If you have any concerns about our products, you can contact us on ProductSafety@springernature.com

In case Publisher is established outside the EU, the EU authorized representative is:

Springer Nature Customer Service Center GmbH
Europaplatz 3
69115 Heidelberg, Germany

FSC
www.fsc.org
MIX
Papier | Fördert
gute Waldnutzung
FSC® C083411

Kolibri 360 GmbH
Ferdinand-Jühlke-Straße 7
99095 Erfurt, Deutschland
produktsicherheit@kolibri360.de